Romantic Science

SUNY series, Studies in the Long Nineteenth Century
Pamela K. Gilbert—editor

Romantic Science

The Literary Forms of Natural History

Noah Heringman, editor

STATE UNIVERSITY OF NEW YORK PRESS

"Lyrical Strategies, Didactic Intent: Reading the Kitchen Garden Manual" is an adapted version of chapters 7 and 8 appearing in Rachel Crawford, *Poetry, Enclosure, and the Vernacular Landscape, 1700–1803*, 2002. © Rachel Crawford, reproduced with permission of Cambridge University Press.

Published by
State University of New York Press, Albany

For information, address State University of New York Press,
90 State Street, Suite 700, Albany, NY 12207

Production by Kelli Williams
Marketing by Jennifer Giovani

Library of Congress Cataloging-in-Publication Data

Romantic science : the literary forms of natural history / Noah Heringman, editor.
    p. cm. — (SUNY series, studies in the long nineteenth century)
    Includes bibliographical references and index.
    ISBN 0-7914-5701-X (alk. paper) — ISBN 0-7914-5702-8 (pbk. : alk. paper)
    1. English literature—19th century—History and criticism.   2. Nature in literature.
  3. Literature and science—Great Britain—History—19th century.   4. Natural history
in literature.   I. Heringman, Noah.   II. Series.

PR468.N3 R66 2003
820.9'36—dc21
                                                                              2002042642

10 9 8 7 6 5 4 3 2 1

Contents

Figures

Cover Image

"Dragon Arum" from Robert Thornton, *The Temple of Flora* (1799–1807).

Facing Page 1

Frontispiece to *Peter's Prophecy* by Thomas Rowlandson (1788)

Chapter 1

Chapter 2

Chapter 5

Chapter 6

Chapter 8

A Note About the Cover

The cover illustration, *The Dragon Arum,* was published in Dr. Robert Thornton's *Temple of Flora,* a work issued in parts between 1798 and 1807 and distinguished by the elaborateness of its engraved plates and accompanying texts as well as its atlas folio size (56 cm by 44 cm, with varying sizes for individual plates). Painted by Peter Henderson and engraved on copperplate by W. Ward, this illustration (Plate 22) is dated December 1, 1801. Together with the other forty colored plates that were published over the nine-year run of *The Temple of Flora, The Dragon Arum* displays some of the engraving techniques refined for large-scale reproduction during the Romantic period, in this instance mezzotint with aquatint added to the last of its three states. The engravings were thus printed in color, then finished by hand, in some instances to offset the degradation of the soft copper with each impression. Identified by its subtitle as a set of illustrations to the Linnaean system, *The Temple of Flora* was at once enormously expensive to produce and buy, a capital investment designed to take advantage of the widespread popularity of illustrated books and magazines dealing with botany and other aspects of natural history. Thornton hoped thereby to make an enormous profit from its publication; instead, he went very nearly bankrupt and was barely saved from immediate financial ruin by a Royal Lottery in which the original paintings of *The Temple of Flora* were the first prize; other prizes included bound copies and loose plates.

Like *The Dragon Arum,* many of the illustrations were of exotic plants. Exotic or native to England and Europe, the flowers in these illustrations are remarkably placed in the foreground of botanical landscapes that extend the visual tradition inaugurated by Maria Sibylla Merian in the early eighteenth century. Whereas Merian's compositions emphasize ecological relations between plants and insects, those commissioned by Thornton are deliberately dramatic, such that even relatively small flowering plants like the "American Cowslip" are made to look larger than the landscape features depicted behind them. This visual aggressiveness is especially marked in its illustration of exotic flowers like the Dragon arum and augmented by Thornton's commentaries. Of the Dragon arum, he declares:

> This extremely foetid poisonous plant will not admit of sober description. Let us therefore personify it. She comes peeping . . . with mischief fraught; from her green covert projects a horrid spear of darkest jet, which she brandishes aloft: issuing from her nostrils flies a noisome vapour infecting the ambient air: her hundred arms are interspersed with white, as in the garments of the inquisition; and on her swollen trunk are observed the speckles of a mighty dragon; her sex is strongly intermingled with the opposite! confusion dire!

Making pointed use of a figurative strategy developed by Linnaeus and extended by Erasmus Darwin in *The Botanic Garden* (1791), Thornton's highly sexualized botanical personifications shamelessly exploit the commercial value of sex, especially kinky sex, in a gothic register. The equally obvious political messages Thornton inserts here (its "garments of the inquisition" make this arum the inevitable exotic adversary to the Protestant English) and elsewhere in *The Temple of Flora* specify another set of claims on the patriotism (and purse) of the British public.

(For further discussion and a list of sources, see Chapter 8, especially note 11.)

Theresa M. Kelley

Acknowledgments

I would like to thank:

The National Endowment for the Humanities for a Long-Term Fellowship at the Huntington Library in 2000–2001, which was instrumental in helping me to complete this collection; and the University of Missouri–Columbia Research Council for a grant supporting the preparation of the manuscript. The Huntington Library staff were especially helpful and courteous. I would also like to thank the editors of *Nineteenth-Century Contexts* and *Critical Inquiry* for allowing me to reprint, respectively, Anne Mellor's and (in somewhat altered form) Lydia Liu's essays. Chapter 7 is an adapted version of chapters 7 and 8 appearing in *Poetry, Enclosure, and the Vernacular Landscape, 1700–1830* (2002). © Rachel Crawford, reproduced with permission of Cambridge University Press.

My contributors, especially Amy King, with whom I hatched the plot for this collection and who coedited with me during the first two years of what came to be a very lengthy project. This collection would not exist without Amy's collaboration, vision, and friendship. Each of the other contributors has contributed much more than just his or her words along the way, and I thank them for their generous collegial spirit.

Alan Bewell, Kevin Gilmartin, David Perkins, Allen Thiher, and especially Jeff Williams, for reading and improving my introduction, and Joe Green, for valuable editorial assistance.

My family, friends, and colleagues, for enabling my perseverance on this project, and James Peltz of State University of New York Press for believing in the project with me.

The adventurous undergraduates in Literature and Science at the University of Missouri–Columbia, 1999–2001, for helping me to understand not only the importance of the subject but also the value of speaking about it in a cogent and accessible way.

Elizabeth Hornbeck, for building not one, but two homes with me to sustain work and life and joy.

Frontispiece to *Peter's Prophecy* by Thomas Rowlandson (1788).

Introduction

The Commerce of Literature and Natural History

NOAH HERINGMAN

Objects and Disciplines

When Wordsworth's *The River Duddon* appeared in 1820, eleven years before Charles Darwin embarked aboard the *Beagle,* the publishers saw an opportunity for direct marketing to readers interested in natural history: the endpapers of Wordsworth's volume of nature poetry advertise "New Works on Botany, Gardening, Agriculture, Geology, Mineralogy, Philosophy, &c. published by Longman," including general works of natural history, theoretical treatises, gardeners' manuals, and even mock-heroic poems on geology such as "King Coal's Levee; or Geological Etiquette."[1] Such vivid illustrations of "the availability of scientific culture" in the Romantic period are still surprising—they suggest that much of what we now call "science" was embedded in a cultural network more established and more vast than what has emerged in the revealing literary scholarship on Darwin.[2] Longman's advertisement documents the most literal form of commerce joining literature and natural history in Wordsworth's time: the publishing industry. The rapid expansion of print culture beginning in the later eighteenth century fueled the circulation of writings famously obsessed with nature, from Romantic poems and scenic tours to theories of the picturesque or the Deluge to the persistent and polymathic genres of natural history. These kinds of writing shared a common readership, and major publishers such as Longman (and authors such as Wordsworth) promoted their long-standing affiliation. At the same time, the increasingly self-conscious establishment of many of these discourses as disciplines—including geology, meteorology, aesthetics—led to new efforts both to join and to distinguish different kinds of objects and modes of cultural production.

This book addresses the mutually constitutive nature of literary and scientific discourses in Britain during the later eighteenth and early nineteenth centuries: the period of what is sometimes called the "second scientific revolution."[3] The contributors analyze the signs of this commerce of literature and natural history, which also operate as currency in more symbolic forms of commerce: applied science, global and imperial exchange, cross-disciplinary appropriations, and other forms of cultural production and reproduction. These signs include a decaying, badly taxidermied moose sent by Thomas Jefferson to the Comte de Buffon in Paris to prove the vitality of American species; a collection of fossils sold to the British Museum and inspiring visions of the earth as a vast fund of knowledge and resources; a shattered oak tree that taints an Edenic vision of British climate and history by registering a violent tempest and a symbolic threat to British naval superiority; and a vegetable—the cucumber—whose sexual potential is neglected by poets only to be exploited by the authors of kitchen garden manuals. These objects evoke the larger economy of ideas and resources both reflected and regulated by the print culture encompassing literature and natural history in late-eighteenth- and early-nineteenth-century Britain.

The economy governing the commerce of literature and natural history is not just the subject matter of this book, but continues to operate in the institutional culture producing the scholarship collected here. The chapters that follow resist constraints of specialization that were just beginning to form in the early industrial era. Though the "economy of nature" pursued by that era's intellectuals seems quite remote from the objects of current scholarship, both phases are chapters in the same story: the story of the modern disciplines. Timothy Lenoir has recently pointed out economic continuities within this story, arguing that "disciplines are political structures that mediate crucially between the political economy and the production of knowledge."[4] Antiquated though it may appear at first glance, the satire in our frontispiece targets both of these modern and postmodern issues: the conflict between disciplines and the connection between knowledge and the market. This image first appeared in 1788 as the frontispiece to a satirical poem by "Peter Pindar" (John Wolcot) addressed to Sir Joseph Banks, the period's most eminent naturalist, who is seen in the cartoon devouring a snake. The verses are "a lengthy attack on Banks for . . . ejecting [natural] philosophy . . . from the [Royal] Society in the petty interests of natural history."[5] Most obviously, the cartoon ridicules the activities associated with natural history and Banks in particular: the collection and classifying of numerous exotic specimens. The verses spell out the implication that natural philosophy—the competing natural science discipline in the late eighteenth century—is the more serious discipline and ought to remain at the center of official scientific activity. In his first decade as president of the Royal Society, Banks had successfully asserted the priority of natural history. His other major office was that of de facto scientific director of Kew, the royal botanical garden, to which he referred in 1787 as "a great botanical exchange house for the empire."[6] This commercialization of botany provides the economic basis for the cartoonist's satire: the problem with natural history (of which botany was one branch) is that it is tainted by ulterior motives. This remains a central objection of many critics of modern science.

Our frontispiece puts into play the literal economics of natural history and its literary context, making it an apt emblem for our investigations of the material culture surrounding natural history in the period. The conflict between disciplines at issue in the 1788 cartoon—and, in its present-day form, underlying the chapters that follow—highlights the importance of the literary context for natural history's political and economic ramifications. The cartoon and the poem it accompanies criticize Banks's commercialization of science partly because it disrupts an established model of science as a literary and philosophical pursuit, enshrined in the title of the Royal Society's profoundly influential journal *Philosophical Transactions*. The evolution of natural history as a commercial sphere is a prominent theme in this volume (see especially chapters 2, 4, 5, and 7). The cartoon also links the proliferation of individual specimens with individual enterprise, a conjunction important to this volume in several ways. On the one hand, Romanticism is traditionally understood as being concerned with the particularity of nature and with individual identity, and it is increasingly understood as being enmeshed with intellectual professionalization. The natural history satirized in Pindar's frontispiece is "Romantic science" in all these senses. The cartoon is skeptical about specialization, as Romanticism is often held to be. Yet Wordsworth's poetry (see chapter 1), like Banks's natural history and the many contemporaneous forms of natural knowledge investigated in this volume, is inextricably a part of the long episodic process of specialization leading to the formation of modern disciplines.

Though "natural history" now has several new and different meanings, it remains a familiar category. "Natural philosophy," despite the misgivings of Peter Pindar and the cartoonist, was finally eclipsed by the end of the nineteenth century. Both have been superseded by a structure of modern, specialized disciplines, but "natural history," however modified, persists as a broad rubric for the knowledge and practice of amateur naturalists and for the miscellaneous specimens collected in museums. The conflict between natural history and natural philosophy is still with us in the view that natural history is not proper science in terms of institutional prestige and explanatory power. The cartoon is prophetic in this sense: although Banks contributed to a slow, far-reaching process of specialization that gradually produced various earth, plant, and animal sciences in place of the three traditional "branches" of natural history, the name "natural history" stuck to a set of ostensibly unrigorous cultural practices. This volume considers natural history as a literary practice during a transitional phase, along with numerous other social and material practices for the transmission of natural knowledge. Many of these

evolved into newly specialized scientific practices, though natural history also retained some of its key features, including its generic identity as literature. The
length of this transition has produced a number of contradictory explanations in
the history of science and culture. This is why "natural history . . . seems to have
come to an end so often," as James Secord puts it.[7]

The image of Banks's dinner for naturalists, like the chapters in this book, pin-
points an important local shift within the long, slow process of intellectual
specialization. This process spans more than three hundred years, from the initial
conflict between natural history and natural philosophy to the present-day proliferation both of ever-more specialized disciplines such as "nanoscience" (now a
separate Ph.D. field at the State University of New York–Albany, with other programs pending) and of interdisciplinary initiatives such as the present volume. The
crucial distinction between natural history and natural philosophy in the Romantic period was that the former called for fieldwork, the component that kept natural
knowledge accessible to a generalist public, both in the field and on the printed
page. This distinction was crucial partly because it permitted the synthesis between
theory and fieldwork that produced many of the modern disciplines. Historians of
geology, for example, use the foundation of the Geological Society in 1807 as their
touchstone for discipline formation; much of natural philosophy, by contrast,
had been codified by Newton, and even chemistry, though still solidifying in the
era of Humphry Davy, was already characterized by a sharper cleavage between
practitioners and the public.[8] Both the empirical and the literary components of
natural history can be traced to Pliny and the Renaissance revival of natural history. But as natural history was revived, the Copernican and associated
"revolutions," culminating in Newtonianism, created a competing disciplinary
structure out of natural philosophy by grounding it in "mathematical principles"
and formulating a distinct formal language for physical and astronomical knowledge. The process of discipline formation, then, can be traced at least as far as the
publication of the *Principia* in 1686. *The Origin of Species* (1859), however, is just
as commonly taken as a landmark of discipline formation and the divergence of literature and science. This amorphous chronology leads to the massive
contradictions alluded to by Secord. Mark Greenberg, focusing on Newtonianism,
argues that poets were at work as early as the 1720s "appropriating" the power and
prestige of a science they already recognized as the stronger of "the two emerging
institutions."[9] Gillian Beer, on the other hand, just as confidently refers to Darwin's idiom as "literary . . . discourse," explaining that in Darwin's time "scientists
themselves in their texts drew openly upon literary, historical and philosophical
material as part of their argument."[10]

Romantic Science proceeds from two contradictory premises: self-consciously
interdisciplinary method and historically specific scholarly practice. It aims to
historicize the disciplinary boundaries between literature and science not only in

the content but in the form of the chapters, each of which reimagines and re-draws those boundaries in relation to their current forms. Alongside this intellectual revisionism, the collection's distinctive focus on the Romantic period allows for concrete and specific examinations of the origins of our own intellectual landscape. Contradictions inevitably attend this intervention in the historical and method-ological space between the above accounts of Newtonianism and "Darwin's Plots." Sensitivity to the historical uniqueness of Romantic natural history (à la Beer) is conditioned by the inexorable triumph of Newtonianism (à la Greenberg). While natural philosophy, as we have seen, gradually lost its currency, its main areas con-tinue to stand atop the hierarchy of disciplines. As Secord puts it in *Cultures of Natural History*, "histories of the life and earth sciences continue to be dominated by fields . . . that claim a theoretical generality akin to that of physics" (458). Margery Arent Safir brings out the implicit connection between this domination and the "weaker" institutional position of literature: "are authors of fiction seeking to share in the power and authority contemporary society accords science?"[11] The context for her question is a post-disciplinary climate that she sees as a potential site for the return of an educated public pursuing encyclopedic knowledge "in the fashion of Buffon" (11). In keeping with Safir's call for a "more comprehensive and more inclusive interdisciplinary practice" among academics (18), this volume com-bines and considers many different disciplines, providing points of access for many different readers.

A post-disciplinary climate, however, cannot be created without a careful ex-amination of cultural history. Most of the contributors to this volume are professional scholars of British Romanticism, one of several disciplines currently en-gaged in reevaluating and (as it were) historicizing historicism. The volume itself is published by a university press series devoted to the "Long Nineteenth Century." The title invokes recent efforts to reconsider the use of centuries as the main orga-nizing principle of scholarship in the humanities, a legacy of the professionalization of that scholarship in the nineteenth century. One such effort is documented in the *PMLA* forum "Who's Carving Up the Nineteenth Century?" (October 2001), which features Romanticists reflecting on the anxieties associated with this period that is not a century, wedged between the ever longer eighteenth century and the Victorian period. One of these essays offers an anonymous epigraph that raises com-pelling questions about the carving up of centuries in the humanities: "If you don't say to them that . . . the legacy of 19th century historicism was 20th century aca-demic territoriality, that the artifice of centuries is only useful as a foil against which to pose searching queries about poetry, cultural expression, textual practice etc. . . . then who will?"[12] In response to such questions, this volume offers a purposive his-toricism, a focused cultural history that acknowledges its heuristic relation to larger questions concerning the compartmentalization of culture. Each of the essays will suggest differences between a pre-disciplinary and a post-disciplinary climate, which cannot be imagined as a return to the culture of natural history. Nor will such a climate be produced by a return to unified standards of causal explanation—as Safir also advocates—according to which a scientific theory is relevant to a work only when its author is known to have studied the theory; nor by a return to tried and true definitions of Romanticism as a small body of poetry produced by a handful of geniuses. All these are disciplinary constructions, merely reflecting the extent to which culture has been disciplined.[13]

Romantic Natural History

One of the virtues of natural history, then as now, is that it helps to unite local scholarly efforts in literature and science. The scientific backgrounds of specific texts are important in themselves and as distinct moments of scientific culture, but they are also jointly important as evidence of the shared culture of "letters" and of the epistemological claims of literary projects to explain the natural world. Such claims reflect the historical importance of aesthetic forms for the transmission of scientific concepts—the transmission that ultimately makes possible the division of the scientific disciplines from literary culture. These historical circumstances inform two of the distinctive contributions of *Romantic Science*: its precise historical focus on the Romantic period and the resulting effort to redeploy categories of literary history, such as "Romanticism," within the broad new context of cultural history emerging from the work of literary scholars and historians of science. Most broadly, the chapters of this book are united by their shared focus on English literature and natural history between Linnaeus and Darwin and the common issues that arise from the particular convergences of literary and scientific culture during that period.[14] The term "natural history" itself, as it was used in the period, designates a subset of the practices that make up this scientific culture, creating a focus distinct from that of synoptic collections on literature and science as well as other scholarship on Romanticism and "the sciences." Although most of our contributors work primarily on English literature, the collection also reflects the increasingly global reach of the period's natural history.

Natural history, as a literary practice, shares the style and resources of other literary forms, while scientific epistemology and the cultural imperative of "improvement" with which it is associated also determine the orientation of fiction and poetry in the period. The increasing and simultaneous popularity of natural history, poetry, and the novel in the late eighteenth and early nineteenth centuries created audiences predisposed to synthesize these now divorced forms in their reading and their actual experience of the natural world. The pervasiveness of natural history, though equaled by other nonspecialist discourses, has a special importance because it ultimately precipitated the epochal rupture between "literature" and "science." By stressing the literary affiliations of natural history, this collection also contributes to the dialogue between two sometimes competing disciplinary structures of our own intellectual landscape, literary criticism and history of science. While "science studies" can currently be subsumed under cultural studies, this tends to favor the study of twentieth- (and twenty-first-) century culture and has provided a target for polemics against the humanities.[15] Entering the field as a historical form of science studies, this book also reflects on the disciplinary boundaries remaining between literary study and history of science, an epiphenomenon of the more fundamental division between humanities and sciences that necessitates our recovery of the interdisciplinary or pre-disciplinary space of natural history. On the one hand, many of the chapters focus on familiar texts, such as *Frankenstein, Prometheus Unbound,* and *Robinson Crusoe,* bringing the cultural context of science to bear on widely recognizable objects of literary study. On the other hand, each of the chapters also draws on and responds to the work of historians of science, contributing to the ongoing effort at cross-fertilization between the two methodologies.

The generic and cultural richness of Romantic natural history holds particular rewards for literary analysis. The genres of natural history explored in *Romantic Science* include numerous verse forms, fictional and nonfictional narratives, journals, letters, public lectures, and kitchen garden manuals, among others. This approach continues a project laid out by the cultural historian Ludmilla Jordanova: "History of science can learn much from the methods of literary criticism, particularly in textual analysis. Treating scientific writings as literary texts involves, for example, asking questions about genre, about the relationship between reader and writer, about the use of linguistic devices such as metaphor, simile, and personification, about what is *not* being said."[16] Other studies have proven this principle, but it is especially apt here because of the incipient specialization and broad literary purchase of Romantic natural history.[17]

There are numerous other reasons for a volume on literature and science to focus on natural history of the Romantic period. Such a project benefits from the fruitful paradox implied by "Romantic science," a prominent phrase in much recent work by Romanticists—though hitherto unclaimed as a book title.[18] It is a paradox because hostility to science is still a major feature of most popular definitions of Romanticism, which draw on a handful of famous quotations such as Wordsworth's "we murder to dissect." As a consequence, the deeply scientific concerns of the period often reveal themselves in surprising and illuminating ways. Jordanova settles on a largely overlapping period, 1760–1820, in *Nature Displayed* (1999) but registers uneasiness with the term "romantic" as a name for this period because it denotes, for her, an artistic "style."[19] Romanticists themselves, however, have been engaged since the 1980s in articulating a shift from stylistic to historical definitions of this term. Such combinations as "Romantic natural history" are part of the effort to map this terrain historically, an effort overlapping to a large extent with Jordanova's focus on thinkers who are "products of the Enlightenment," "largely professionals and middle-class," and "worshippers of Nature" (*ND* 6)—a phrase borrowed from Wordsworth's "Tintern Abbey." German Romanticism has been especially fertile ground for work on literature and science because major figures including Goethe, Novalis, and Alexander von Humboldt had scientific training and began their careers as state inspectors of mines.[20]

The period's publishing industry produced an unprecedented amount of natural knowledge for a general audience, and this abundance in turn provided the point of departure for increased specialization and professionalization. Elinor Shaffer identifies the intellectual engine of this abundance: "the late eighteenth century and early nineteenth century was [*sic*] especially fertile in attempts to synthesize theoretical material drawn from more than one subject, and to find means to come to terms with the increasing domination of scientific modes" (*TTC* 7).[21] One reason for the period's fecundity is the emphasis on "Nature" investigated by previous generations of Romanticists—as Jordanova points out, "'Nature' included such a multitude of phenomena to be described, marveled at and investigated, that it was inevitable that there should be considerable common ground between the areas we designate as science and literature respectively" (*LN* 23). The growth of literary interest in nature, however, precipitated new divisions of knowledge. The breadth of Romantic natural history was itself conducive to specialization: in the course of our period, its three traditional branches—geology, botany, and zoology—moved rapidly toward the disciplinary status achieved much earlier by the physical sciences, as detailed above. A burgeoning print culture fueled both breadth and specialization. "Far-reaching changes . . . in the publishing business, and also in education and in newspapers," as Jordanova shows, "brought knowledge about science and medicine to a largely non-specialist, though predominantly middle-class, public" (24).[22]

From the beginning, specialist discourses have relied on the literary as a category against which the scientific may be set off. As George Levine puts it, "[S]cience established itself professionally in England in part by rejecting literature—at least those excesses of literature that seemed to the Royal Society to corrupt thought."[23] Writing in 1684, Thomas Burnet adopts such a rhetorical stance when he distinguishes the "Oratour's" view of the earth as "a beautiful and regular globe" from the "impartial" prose of the "Philosopher" who recognizes it as "a broken and confus'd heap of bodies."[24] John Playfair, however, writing in 1811, rejects Burnet's and other earlier theories of the earth as "a species of mental derangement, in which the patient raved continually of comets, deluges, volcanos and earthquakes; or talked of reclaiming the great wastes of the chaos, and converting them into a terraqueous and habitable globe."[25] This rejection is typical—in the Romantic period some naturalists began to conflate earlier natural history with "literature" in Levine's sense. Although Playfair attempted to distinguish his own self-consciously modern geology from older natural history, Erasmus Darwin had insisted only twenty years before on the importance of obsolete or fanciful theories even in the rigorous precincts of natural philosophy: "[E]xtravagant theories . . . in those parts of philosophy where our knowledge is yet imperfect, are not without their use; as they encourage the execution of laborious experiments, or the investigation of ingenious deductions, to confirm or refute them."[26] Literary qualities—whether embraced for their imaginative "extravagance" or discarded as "raving" excess—were thus essential to specialization and professionalization in numerous areas we now see as strictly scientific. A generally informed readership was not the least of these literary requirements, as the same print market served the professionalization of both literature and science. Theresa Kelley—to cite an example from chapter 8 of this volume—shows that representations of flowers in the writings of Charlotte Smith and John Clare depend on readers' familiarity with increasingly technical botanical works such as Sowerby and Smith's *English Botany*.

"Natural history" is a problematic term partly because it persists as a disciplinary category from the seventeenth century up to *The Origin of Species* and beyond. The phase between Linnaeus and Charles Darwin, however, is a very fertile and important period for natural history: as poets and novelists embrace more strongly than ever before the literary tradition it represents, self-consciously modern and increasingly professionalized naturalists repudiate that literary tradition as empirically unsound and seek to appropriate the term for the emerging sphere of modern science. At the same time, especially toward the end of the period, amateur naturalists seek to reclaim the space of natural history in the name of folk traditions, a point raised in two of the chapters on botany. *Romantic Science* includes a whole section on botany partly because it serves as a bellwether for the transitional phase that focuses the volume: the beginning of this period is strongly marked by the movement to adopt Linnaean botany in England, which begins at mid-century and moves through a set of distinctly literary controversies—some already well-known, others newly elaborated in these chapters (7, 8, and 9). The overlapping chronologies of botany and Romanticism provide a unifying framework for the various other subjects explored here, such as mineralogy and biogeography, and the science and literature evolving from them.

By situating this volume on the border between the cultural study of science and literary scholarship, we aim to reach teachers and students of English literature as well as historians of science and other readers interested in cultural history.[27] The historically specific combination of intellectual breadth with depth and variety of content—as an unprecedented volume of observations of natural phenomena streamed into imperial Britain, and into print—will be evident in the range of subjects treated in these essays. But the volume's historical focus also allows them collectively to demonstrate the importance of the period for the formation of humanistic and scientific disciplines. During the crucial period between Linnaeus and Darwin, the professionalization that would drive the disciplines apart was only just becoming viable, and literary texts provided the cultural currency required for natural history to become an increasingly middle-class avocation.

As a collection of essays by several hands, this volume does not strive for methodological unity. The project as a whole is loosely affiliated with the paradigms of public culture that have been so prominent in the work of scholars studying Romantic literature, Enlightenment science, and the surprisingly extensive area that overlaps these two apparently opposite rubrics.[28] Much of this work is still indebted to the public sphere paradigm of Jürgen Habermas, though the concept of public culture encompasses an increasing range of new possibilities, as this volume helps to illustrate.[29] Many forms of natural history have particular claims to public sphere functionality because they are not only objects of taste but can also be materially serviceable in social advancement (see, for example, chapters 1, 2, and 7 of this volume). Implicitly complicating Habermas's bifurcation of the bourgeois public sphere—in which "women and dependents are practically and juridically excluded from the political public sphere, even though female readers . . . often participate more fully in the literary public than the property owners themselves"—*Romantic Science* also highlights numerous cases in which the literary practice of natural history facilitates women's *political* participation in cultural life.[30]

However, this focus on issues susceptible to Habermasian analysis, such as the political ramifications of a certain branch of print culture and the discursive processes of discipline formation, points more to the intellectual currency of these issues than to the specific influence of Habermas. In chapter 8, for instance, Theresa Kelley argues that John Clare and Charlotte Smith manipulate botanical taxonomy to "gain a strong poetic purchase on one of Romanticism's material cultures." The metaphor of "purchase" nicely brings out one sense of the cultural *commerce* of literature and natural history. Just as Clare and Smith participate intellectually and commercially—here in the sense of competition for overlapping readerships—in the public discourse on taxonomy, Mary Shelley participates in the intellectual milieu surrounding William Lawrence, her personal physician, whose lectures (published as *Lectures on . . . the Natural History of Man*) were attended by Percy Shelley. Her participation in this discourse gains political force, in Anne Mellor's reading (chapter 6), inasmuch as *Frankenstein* scientifically emphasizes racial difference, in dialogue with Lawrence and other race theorists, ultimately for the purpose of promoting racial harmony. Stuart Peterfreund (chapter 3) argues that Gilbert White's *The Natural History of Selborne* (1789) "reveals a symbolic register in terms of which meteorological catastrophes stand for the manifestations of sociopolitical change threatening the established order of England," and he explains how this political dimension is tied both to the volume's initial commercial failure and its tremendous eventual success in the 1820s. Alan Bewell (chapter 4) offers a case study of bic-geography—perhaps the most aggressively colonial form of natural history—to illuminate Thomas Jefferson's advocacy of natural history as a public and political discourse in the early years of the Republic. Catherine Ross (chapter 1) explicitly draws on Habermas, and more particularly on the sociology of the professions, to explain the senses in which Wordsworth and Davy were "rivals in the public sphere." Lydia Liu's chapter (5) may be especially relevant here because she bridges the distance between the early eighteenth century and the modern world—one of the gaps in public sphere theory—as she traces the eighteenth- and nineteenth-century reception history of *Robinson Crusoe* (1719) in terms of prolonged scientific and commercial developments anticipated by the novel.

Introducing the Chapters

This book is in three parts: the first three chapters delineate some of the boundaries of natural history; the next group of three addresses the global reach of natural history; and the last three chapters concentrate on the representative story of botany, as I have mentioned. The first part addresses boundaries that are discursive rather than geographic, boundaries negotiated to some degree in all the chapters. The middle part represents another core concern of the book in its focus on the transcontinental movement of natural history—a legacy of the "republic of letters"—and its global claim on climate (see especially chapter 4), resources (chapter 5), and cultures (chapter 6). While these three chapters deal explicitly with the claim of scientific theory on conditions obtaining on opposing parts of the globe, the cross-cultural transmission of natural history is important in the other parts as well. Theresa Kelley and Rachel Crawford (in part 3) address the "vernacularization" of Linnaean and other botanical systems, for example, while my own chapter (in part 1) involves the concept of the global and its geological origins. Geology, like meteorology and chemistry—the other fields addressed in part 1—was at an active stage of discipline formation during the period covered by this book. Shelley and White (chapters 2 and 3) self-consciously blend archaic and modern concepts from natural history, a move characteristic of the transitional phase of discipline formation. Humphry Davy, in dialogue with Wordsworth (chapter 1), assertively claims a place for chemistry outside and morally equal to the literary culture surrounding natural history. Botany—as illustrated by the last part of this book—arguably maintained its proto-disciplinary complexity longer than the other branches of natural history: it was in the vanguard of all modern life science as the original site of Linnaean systematization, beginning midway through the eighteenth century, but as late as 1823 (chapter 8) English botany remained a healthy tangle of competing systems. The resulting debates retained their political currency (chapters 7 and 9) until even later.

Chapter 1 marks the parameters of our main focus on literary natural history by analyzing the "professional rivalry" between Wordsworth and Davy. Catherine Ross demonstrates a rivalry that is surprising to us in the era of "two cultures" because it was based on close affinities in philosophy and rhetorical style, as well as class background. Applying public sphere theory and the sociology of the professions to this relationship, Ross foregrounds the questions of disciplinarity and polymathism that are at issue throughout the volume. Natural history was the territory contested between the rivals, in the sense that Wordsworth and Davy created distinct professional identities by articulating mutually exclusive claims on "Nature," thus removing it from the shared territory of natural history. Davy instead champions natural philosophy as the forum of scientific modernity, appropriating the standards of accessibility that came from natural history and from Wordsworth, who also tried—with less success—to create a career as a public intellectual using the "real language of men." In chapter 2, I analyze the paradigms of earth history shared between works of Percy Shelley and William Smith, an economic geologist whose work was so accessible that it was dismissed (as well as plagiarized) by the geological establishment and thus failed to produce professional success. In the context of Smith's maps and writings (1815–1817), Shelley's images of rocks as a natural archive in *Prometheus Unbound* (1819) form a crucial chapter in the emergence of the "rock record" as a modern geological paradigm. The culturally pervasive metaphor of the archive is central to both, showing the origins of the "rock record" in applied mineralogy and poetry, two areas of natural history *outside* the disciplinary boundary being erected at the same time by the Geological Society. Turning to the earliest English form of geology and cosmology, chapter 3 examines the discourse of physico-theology and its role in the Romantic natural history of Gilbert White. Stuart Peterfreund argues that White's *The Natural History of Selborne*, drawing on Milton as well as physico-theology, infuses local natural history with hints of global catastrophe. Focusing on apocalyptic allusions to climate change, Peterfreund shows that White's Edenic vision unconsciously registers historical conflict through its dependence on John Ray and William Derham—a dependence which, ironically, limited his popularity at a time when this scriptural science was being condemned as old-fashioned.

Chapter 4 also focuses on climate change, but in the modern context of biogeography, which connects local weather to global climate through the discourse of colonialism, rather than biblical typology. If White's and Wordsworth's popularity was much delayed for reasons relating to their lack of scientific modernity—as Peterfreund and Ross, respectively, suggest—then the modernity associated with colonial biogeography seems to have contributed to the more immediate popularity of Buffon's *Natural History* (1749–1804), for Bewell a major source for the politics of climate on both sides of the Atlantic. At a time when, as

Bewell argues, "climate functioned as a primary category of cultural and political analysis," data on the American climate seemed to justify both republican government and the human capacity to transform physical environments. Bewell's essay analyzes the late-eighteenth-century debate concerning the quality of the New World as a biological habitat, drawing on Jefferson's correspondence as well as the poetry of Keats and Blake to establish a connection between republicanism and climate change. Continuing the investigation of the global effects of natural history, chapter 5 examines British and European anxieties over the circulation and reproduction of Asian prestige commodities. Though Lydia Liu's essay considers our period only briefly in its broad historical contextualization of *Robinson Crusoe*, it is the period in which the scientific and commercial narratives about porcelain, adumbrated in the novel, come together. The nexus of literature and natural history in this context is material culture, specifically the European project of reproducing "true" porcelain, carried on in the sphere of British letters from *Robinson Crusoe* to the *Philosophical Transactions of the Royal Society* to the correspondence of Erasmus Darwin and Josiah Wedgwood. Historicizing the novel's vaunted realism, Liu points out its repressed traces of "science fiction," a repression amplified over the centuries in readings of the novel and in geological experiments that repress the cultural context of Chinese porcelain. The element of science fiction allows her to isolate moments in the novel's reception history—from Rousseau's reading to Jules Verne's and beyond—that correspond to scientific and commercial developments the novel merely anticipated.

In Defoe's and Réaumur's accounts of trying to replicate (or replace) Chinese porcelain, Liu identifies twin legacies that are important for the commerce of literature and natural history in the Romantic period. The multiple senses of "China" in Liu's essay come into play in Wedgwood's commercial project and a section of Darwin's *The Botanic Garden* (1791) devoted to it, in the experiments of James Hall to prove the theories of James Hutton, and even in the rise of *jardin anglo-chinois* on the Continent, with its allegorized tableaux of science and commerce. Anne Mellor (chapter 6) brings out a more general racial anxiety accompanying Asian–European connections made within and between Romantic literature and science. In "*Frankenstein*, Racial Science, and the 'Yellow Peril,'" Mellor shows that the racial identity of Frankenstein's creature is historically conditioned by another body of continental scientific discourse with global claims, physical and medical anthropology. She argues that the creature's characteristics conform to Asian racial types defined by the natural philosopher J. F. Blumenbach. Blumenbach's foundational classification of races was expanded in *Lectures . . . on the Natural History of Man* by the Shelleys' physician William Lawrence, with whom they participated in public and private scientific discussion. The same classification later fueled the racist imagery of a "yellow peril," which Shelley's novel, Mellor argues, anticipates and resists.

Romantic Science concludes with three chapters examining the literary applications of botany, while also continuing Peterfreund's and Bewell's concern with the practice of local natural history. Rachel Crawford's contribution presents kitchen garden manuals as one of the largely unexamined aspects of natural history and the most popular "science" of the eighteenth and early nineteenth centuries. Arguing that kitchen gardens combine didactic purpose with lyric strategies well before the lyric turn which we associate with Romanticism, chapter 7 examines four convergences between the space of the kitchen gardens and that of minor lyric forms, vividly illustrating the interchange of literature and natural history. In "Romantic Exemplarity: Botany and 'Material' Culture" (chapter 8), Theresa Kelley compares the ways in which Charlotte Smith and John Clare use botanical figures and botanical practice to present taxonomy as a vehicle and figure for Romantic dissent. For both poets, botany provides a material basis for specifying the tropological work of individual difference within a community or public. Also looking at botanical taxonomy, Amy King argues in the final chapter of this book that Elizabeth Gaskell draws attention to her working-class characters' powers of classification in order to suggest a dual program of reform. Natural history appears as both a working-class intellectual program merging tradition and enlightenment and an instrument for correcting the philosophically false classification of human beings as belonging to different "classes."

The historical "bookends" of this collection, Liu's chapter on *Robinson Crusoe* and its reception history and King's on Gaskell's *Mary Barton* (1848), highlight the persistence of taxonomy as a major form of currency in the commerce of literature and natural history. *Robinson Crusoe*'s account of earthenware vessels anticipates the increasing importance of such terms as *kaolin* and *petuntse* in the scientific effort to domesticate these substances and apply their mineralogy, emptied of cultural content, to "theories of the earth" in the later eighteenth century. Gaskell, on the other hand, reaches back to the natural history of the eighteenth century and before to support natural history's claim to the status of folk culture. Theresa Kelley, too, takes up the centuries-long contest between proponents of vernacular botanical names and "natural" classification and advocates of binomial nomenclature and other totalizing systems. Charlotte Smith's resistance to such systems, as well as John Clare's (partly inspired by Smith's), has surprising political ramifications, which in turn supports the notion of political efficacy that Gaskell imagines for natural history. The informed reverence for the natural world in Gaskell's novel— even or especially as a nostalgic counterpoint to the grim realities of Manchester—is an obvious legacy of Romanticism. Defoe's erasure of Chinese porcelain taxonomy provides a less traditional, but equally important, connection with Romanticism by introducing the global and colonial dimensions essential to this volume's elaboration of "Romantic natural history" as a cultural category. The cultural and gender politics arising from issues of taxonomy return us to one of the central concerns of this book: the public sphere functionality of Romantic natural history, especially in relation to women.

The taxonomic argument in Rachel Crawford's chapter (7) is literary rather than natural, extending the definition of what constitutes a literary text. Rather than viewing literature through the discourse of natural history, as is traditional, her approach examines an overlooked aspect of natural history through the lens of poetics. This apparent inversion of relationships highlights a central method of this collection: making the forms of natural history the object of literary analysis. In Crawford's essay, for example, the ideology of social and sexual productivity finds a forum both in the lyric and the kitchen garden manual, while in Liu's essay the "poetics of colonial disavowal" equally informs readerly responses to a novel and the scientific and commercial quest for "true porcelain." The purpose of this method is not to prove once again that natural history is a cultural artifact, but to show it as a form meaningfully cognate with other literary forms of the period. To this end, *Romantic Science* considers works of natural history side by side with works of "imaginative literature," illuminating the multitude of concrete and specific relationships that exist within and among these works, and establishing an important historical dimension of both literature and natural history as interlocking cultural practices.

Notes

1. John Wyatt's *Wordsworth and the Geologists* (Cambridge: Cambridge UP, 1995), 64, first drew my attention to these advertisements. I have since examined the Huntington Library copy of *The River Duddon, a series of Sonnets: Vaudracour & Julia: and other Poems. To which is annexed, a topographical description of the Country of the Lakes . . .* (London: Longman, Hurst, and Rees, 1820). The two comic poems advertised by Longman are the fourth edition (!) of "King Coal's Levee; or Geological Etiquette, with Explanatory Notes; and the Council of Metals. To which is added, Baron Basalt's Tour" and "A Geological Primer, in Verse; with a Poetical Geognosy," both priced at four shillings.

2. Major studies of this later period include Gillian Beer, *Darwin's Plots: Evolutionary Narrative in Darwin, George Eliot, and Nineteenth-Century Fiction* (London: Routledge, 1983); and George Levine, *Darwin and the Novelists: Patterns of Science in Victorian Fiction* (Cambridge, MA: Harvard UP, 1988). "The availability of scientific culture" is John Wyatt's phrase (64).

3. See, for example, A. Cunningham and N. Jardine, eds., *Romanticism and the Sciences* (Cambridge: Cambridge UP, 1990), xix–xx; and Michel Foucault, *The Order of Things: An Archaeology of the Human Sciences* (1966; New York: Random House, 1970), xxiii.

4. *Instituting Science: The Cultural Production of Scientific Disciplines* (Stanford: Stanford UP, 1997), 47. Lenoir goes on to argue that "disciplines do not have single originary sources, but are more appropriately grasped as interactive system effects. The idea of an economy best captures this sort of dynamic" (52).

5. M. D. George, *Catalogue of Political and Personal Satires Preserved in the Department of Prints and Drawings in the British Museum*, 9 vols. (London: British Museum, 1949), 6:554–55, item 7431. The Wolcot poem is entitled *Peter's Prophecy . . . or, an important Epistle to Sir Joseph Banks* (London: G. Kearsley, 1788). Wolcot chastises Banks for using science as an instrument of power and for "vaunt[ing] on his great acquaintance with vegetables and monkeys" ("Argument").

6. Ray Desmond, *Kew: The History of the Royal Botanic Gardens* (London: Harvill, 1995), 126, quoting Banks's letter to H. Dundas of June 15, 1787. Cf. Desmond's remarks on botany and global commerce (91).

7. N. Jardine, J. Secord, and E. C. Spary, eds., *Cultures of Natural History* (Cambridge: Cambridge UP, 1996), 449.

8. On the last point, see Jan Golinski, *Science as Public Culture: Chemistry and Enlightenment in Britain, 1760–1820* (Cambridge: Cambridge UP, 1992). Developments comparable to the foundation of the Geological Society, in the areas of botany and zoology, might include, respectively, the formation of the Linnaean Society in 1788 and the publication in 1812 of Georges Cuvier's *Recherches sur les ossemens fossiles de quadrupèdes*. The moment of discipline formation is not as clearly demarcated in zoology and botany, which might be said to coalesce into modern biology after 1828 (from the discovery of the cell) or after Darwin.

9. "Eighteenth-Century Poetry Represents Moments of Scientific Discovery," *Literature and Science: Theory and Practice*, ed. Stuart Peterfreund (Boston: Northeastern UP, 1990), 115–37, here 120. Cf. Robert Markley, *Fallen Languages: Crises of Representation in Newtonian England, 1660–1740* (Ithaca: Cornell UP, 1993); and Larry Stewart, *The Rise of Public Science: Rhetoric, Technology, and Natural Philosophy in Newtonian Britain, 1660–1750* (Cambridge: Cambridge UP, 1992).

10. *Darwin's Plots* 7.

11. *Melancholies of Knowledge: Literature in the Age of Science* (Albany: State U of New York P, 1999), 18. Cited parenthetically in the text hereafter. The program described in what follows is informed by suggestive readings of both academic and general intellectual culture. The "upscale popularization" of science Safir delineates (7) has also unexpectedly impacted this volume, in the form of a best-selling new biography of William Smith, a geologist who is the subject of chapter 2.

12. Jerome McGann, "Who's Carving Up the Nineteenth Century," *PMLA* 116.5 (October 2001): 1415–21, here 1415.

13. Or, in Lenoir's somewhat melodramatic formulation: "'The Discipline' as such does not exist; it is at best an abstraction formed in the service of a disciplinary program" (*Instituting Science* 71). All three contributors to the *PMLA* forum seem at least somewhat nostalgic for a more limited canon of Romantic texts; Charles Rzepka, for example, believes that only "a handful of authors," however shifting, will comprise Romanticism's enduring "gold standard" (1429–30). For Safir's advocacy of causal explanation, see *Melancholies* 20.

14. One of our departures from traditional literary history will be obvious from the table of contents, which highlights two novels—*Robinson Crusoe* and *Mary Barton*—that fall outside the Romantic period. This deviation, and the use of Linnaeus and Darwin as

markers for an alternative chronology, are addressed in the section of chapter summaries and the paragraphs immediately preceding it.

15. I certainly do not mean to endorse such polemics, including those in John Brockman's *The Third Culture* (New York: Simon and Schuster, 1995) and the literature attending the "Sokal hoax." However, the cultural study of earlier science is not as susceptible to the same charges of relativism and intellectual fraud simply because the nature of disciplinary boundaries in that period differs so radically from our own. On the other hand, the digital age (as in the work of N. Katherine Hayles) has been a major focus of work on literature and science because it is creating a new transformation of disciplinary boundaries.

16. *Languages of Nature: Critical Essays on Science and Literature* (London: Free Association Books, 1986), 20. Cited parenthetically in the text hereafter as *LN*. Subsequent abbreviations for parenthetical citation are given in square brackets following the title of the work.

17. In a more recent book, Jordanova comments that the period's "richness is only just beginning to be appreciated—it demands a historiography which foregrounds cultural complexities." *Nature Displayed: Gender, Science, and Medicine, 1760–1820* [*ND*] (London: Longman, 1999), 16.

18. The program of the 2000 MLA Convention featured a dozen papers dealing with some aspect of science and Romanticism, including a panel entitled "Romantic Science."

19. *ND* 6. The textuality of science in this period is borne out in a curious way by the fact that important contributions in history of science have been made by trained Romanticists such as Dennis Dean and Morse Peckham. Dean's two substantial contributions to history of geology are *James Hutton and the History of Geology* (Ithaca: Cornell UP, 1992) and *Gideon Mantell and the Discovery of Dinosaurs* (Cambridge: Cambridge UP, 1999). Peckham, who returned to Romanticism as his main scholarly focus, edited *The Origin of Species: A Variorum Text* (Philadelphia: U of Pennsylvania P, 1959), a Herculean labor (considering Darwin's extensive revisions in many successive editions) still of service to students of Darwin.

20. The most recent literature and science collection contains a substantial section on Romanticism, connecting this aspect of German cultural history with the English context. See Elinor S. Shaffer, ed., *The Third Culture: Literature and Science* [*TTC*] (Berlin: de Gruyter, 1998). (This collection devotes equal attention to "Theoretical Approaches" and "Modernism and Post-Modernism.") Two previous collections, *Languages of Nature* and *Nature Transfigured: Science and Literature, 1700–1900*, ed. John Christie and Sally Shuttleworth (Manchester: Manchester UP, 1989), specify the eighteenth and nineteenth centuries as their terrain, though without any particular focus on Romanticism.

21. An impressive body of scholarship bears out these claims about the period. Coleridge emerged as a paradigm case in the early 1980s with Trevor Levere's *Poetry Realized in Nature: Coleridge and the Sciences* (Cambridge: Cambridge UP, 1981), which has been followed by a continuing stream of scholarship, most recently Nicholas Roe, ed., *Coleridge and the Sciences of Life* (Oxford: Oxford UP, 2001). Recent scholarship has also extended the matrix of Romantic science to further poets and a range of historical issues. Examples include Alan Bewell, *Romanticism and Colonial Disease* (Baltimore: Johns Hopkins UP, 1999); Karl Kroeber, *Ecological Literary Criticism* (New York: Columbia UP, 1994); Stuart Peterfreund,

William Blake in a Newtonian World (Norman: U of Oklahoma P, 1998); Alan Richardson, *British Romanticism and the Science of the Mind* (Cambridge: Cambridge UP, 2001); and Wyatt, *Wordsworth and the Geologists.*

22. Cf. Thomas Broman on the interplay between generalist and specialist publications in the 1790s in "The Habermasian Public Sphere and 'Science *in* the Enlightenment,'" *History of Science* 36 (1998): 123–49, here 133.

23. *One Culture: Essays in Science and Literature* (Madison: U of Wisconsin P, 1987), 11. Cp. Christie and Shuttleworth, *Nature Transfigured* 2–3; see also p. 9 on specialization.

24. Thomas Burnet, *The Sacred Theory of the Earth* (1691; London: Centaur, 1965), 90–91.

25. This account of the history of geology (published in the *Edinburgh Review*) is quoted in Roy Porter's *The Making of Geology* (Cambridge: Cambridge UP, 1977), 1.

26. "Advertisement," *The Botanic Garden* (London: Joseph Johnson, 1791), n.p.

27. The fertile collaboration of historians, literary scholars, and humanist–scientists on other topics in literature and science is well documented in the pages of the journal *Configurations* (1993–), published by the Society for Literature and Science. Eight collections on literature and science have been published since Jordanova's *Languages of Nature* in 1986, five of them between 1989 and 1991—a crucial period in the evolution of scholarly structures for the study of literature and science. This phase of organization is documented in Peterfreund's *Literature and Science* (4) and Elinor Shaffer's introduction to "Literature and Science," a special issue of *Comparative Criticism* (13 [1991]: xiv–xxix). More recently, somewhat diverging views of the "emergence" of science and literature as a field have appeared in Shaffer (*TTC* 5) and Desiree Hellegers, *Handmaid to Divinity: Natural Philosophy, Poetry, and Gender in Seventeenth-Century England* (Norman: U of Oklahoma P, 2000), 3–4.

28. Much reflection on the "case study"—by Romanticists as well as historians of science—has accompanied the shared trend toward specifying historical moments within public culture. See, for example, James Chandler, *England in 1819: The Politics of Literary Culture and the Case of Romantic Historicism* (Chicago: U of Chicago P, 1998), 6; and William Clark, Jan Golinski, and Simon Schaffer, eds., *The Sciences in Enlightened Europe* (Chicago: U of Chicago P, 1999), 29.

29. Habermas develops his paradigm in *Strukturwandel der Öffentlichkeit* (1962; Frankfurt: Suhrkamp, 1990). It has been much debated whether and how Habermas's paradigm applies to England after 1750, as he turns from the detailed case study of what might be abbreviated as the "London coffeehouse setting" of 1680–1730 (92, cp. 107) to a broader view of European history. The bourgeois public sphere depends, he says, on the fictive identity of the "human" and economic roles (121), but he traces the connection to various events in English history spanning a period of 150 years, the latter four-fifths of which is not analyzed in cultural terms. Steven Shapin and Simon Schaffer's contribution in bringing Habermas to bear on the scientific revolution has been highly influential (see *Leviathan and the Air-Pump: Hobbes, Boyle, and the Experimental Life* [Princeton: Princeton UP, 1985]) and was extended to the period 1760–1820 by Golinski in *Science as Public Culture* (5, cp. 176–87). Other important discussions of science and the public sphere in our period include Thomas Broman, "The Habermasian Public Sphere," and

Clark, Golinski, and Schaffer in their introduction to *The Sciences in Enlightened Europe* (esp. 25–26; cp. 39 for Dorinda Outram's more skeptical view).

30. I quote *Strukturwandel* 121 [my translation]. Habermas himself comments, in a lengthy new preface written for the 1990 edition, that his original account does not adequately address exclusions from the public sphere, particularly that of women (18–20). This limitation has provided the opening for a great deal of revision of Habermas's paradigm. See further *Romanticism and Its Publics: A Forum*, ed. Jon Klancher, *Studies in Romanticism* 33 (Winter 1994); Kevin Gilmartin, *Print Politics: The Press and Radical Opposition in Early Nineteenth-Century England* (Cambridge: Cambridge UP, 1996); and Anne Mellor, *Mothers of the Nation: Women's Political Writing in England, 1780–1830* (Bloomington: Indiana UP, 2000). The history of geology—to take one example of science in the public sphere—yields examples of women from across the class spectrum gaining access to geological practice, from the Dorset orphan Mary Anning to Georgiana, Duchess of Devonshire, while the broader political purchase of natural history is illustrated by the periodical debate over James Hutton's *Theory of the Earth* (1788/1795), which became a coded debate about materialism, prompting Erasmus Darwin's allegorical images of geological "revolution" in *The Economy of Vegetation* (1791).

Part I

The Boundaries of Natural History

CHAPTER ONE

"Twin Labourers and Heirs of the Same Hopes"

The Professional Rivalry of Humphry Davy and William Wordsworth

CATHERINE E. ROSS

A relation (always social) determines the terms and not the reverse.

—Michel de Certeau

I

On October 27, 1820, while touring the Continent, William Wordsworth breakfasted with Thomas Moore in Paris. Moore describes this visit in his journal, noting that Wordsworth "talked a good deal so very smugly" about other poets. According to Moore, Wordsworth commented on "Byron's plagiarisms from him" and on the younger poet's "laboured & antithetical sort of declamation." He complained as well about the "bad vulgar English," the "infinite number of clumsy things . . . and commonplace contrivances" in the novels of his friend Walter Scott.[1]

Later in the day Moore called on Lady Davy, the wife of Sir Humphry Davy (1778–1829), the celebrated natural philosopher, who had been acquainted with Wordsworth and Coleridge since he helped edit the second edition of *Lyrical Ballads* in 1800. Moore's journal continues:

We talked of Wordsworth's exceeding high opinion of himself & [Lady Davy] mentioned that one day, in a large party, Wordsworth, without any thing having been previously said that could lead to the subject, called out suddenly from the top of the table to the bottom, in his most epic tone "Davy!" and (on Davy's putting forth his head in awful expectation of what was coming) said "Do you know the reason why I published The White Doe in Quarto?"—"No, what was it?"—"To show the world my own opinion of it." (356)

This episode illustrates that a pattern of social intercourse and professional rivalry, long acknowledged to exist among Romantic writers, also existed among some of those writers and contemporary natural philosophers—specifically between Wordsworth and Davy. The rivalry between Wordsworth and Davy is of particular interest not, as we might expect, because a poet and scientist were at odds, but because these two intellectual standard-bearers had so much in common and were perceived as such in their time. Drawing upon the work of literary scholars and historians of science as well as sociological theories of the public sphere and the professions, this essay explores the Davy/Wordsworth relationship. It examines texts by both men, including selections of their prose, two of Davy's most important Royal Institution lectures, Wordsworth's Preface to *Lyrical Ballads*, and portions of the 1805 *Prelude*. The *Prelude*, in fact, is the text in which Wordsworth specifies a cognate relationship between the Poet and the "Man of Science": "[B]ard and sage," he says, are "[s]ensuous and intellectual . . . twin labourers and heirs of the same hopes."[2] These words and his balanced syntax, with its coordinate conjunctions and parallel structure, suggest that in 1805 Wordsworth understood the relationship between poets and scientists to be one of personal and professional kinship. It is my claim that a social relation of this sort, along with the attention Davy commanded as the Romantic Age's most prominent and passionate Man of Science, helped to "determine the terms" in which Wordsworth constructed some of his most enduring ideas about Romantic poetry and poets. Evidence of Davy's fame in the Romantic public sphere, as well as similarities in the language, rhetoric, aims, and experience (both public and private) of the scientist and the poet, help to clarify why Wordsworth emphasized experiment, expertise, reason, and "system" in his poetics and claimed that poetry is "the breath and finer spirit of all knowledge . . . the impassioned expression which is the countenance of all science."[3] I argue that both poetry and science required an affirming audience, addressed virtually the same polymathic public, and vied for the same jurisdiction (that of philosopher and sage); hence, their similarities became a professional problem. A rivalry ensued that became the catalyst for both groups to initiate the delineation and emphasis of their differences.

What we perceive today—that poets and scientists are radically different kinds of workers—is not a natural, but rather a constructed estrangement that was necessitated by the changing market for the products of these intellectual laborers during the Romantic Age. As Jan Golinski and David Knight have both aptly observed, part of what distinguished Davy's career was his emphasis on the usefulness of natural philosophy.[4] His first attention-getting work, on nitrous oxide, was part of an effort to cure tuberculosis. At the Royal Institution, much of Davy's research proceeded from requests of its landholding sponsors for assistance in improving the

Figure 1.1. Sir Thomas Lawrence, *Sir Humphry Davy,* 1815, Oil on Canvas. Reproduced by permission of the Royal Institution, London, UK/Bridgeman Art Library. *Davy as "Man of Science."*

Figure 1.2. Thomas Lindsay, from a drawing by B. R. Haydon, *William Wordsworth* ("The Brigand"), 1818, Engraving. Reproduced by permission of Dove Cottage, The Wordworth Trust, Grasmere, UK. *Wordworth as "Man of Letters."*

productivity of their estates. Davy worked, for example, on fertilizers, seed germination, livestock nutrition, and methods of tanning. The practical applications of Davy's creative intellectual labor not only lessened pain or improved conditions for English people, but they also created jobs and new wealth. Additionally, when applied to other developing sciences, such as engineering, geology, or metallurgy, the advances in knowledge that Davy and other experimentalists effected helped to change social and economic systems in Britain forever. While many historians of science have examined these social phenomena, the subject of how Davy's labor and his public persona impacted literary thought and cultural production has not yet been fully examined. The subject invites attention, for even as Davy's applied science was making a material difference in society, Davy himself continued to speak of the aims of natural philosophy in a language that was so similar to that of the Romantic poets that many members of their public noted the kinship. Understanding the social, philosophical, and professional relationships of men such as Davy and Wordsworth empowers scholars to read Romantic texts, especially those about the purposes, job, and commerce of writing, with greater precision and insight.

This examination of the relationship between Davy and Wordsworth expands upon the work of critics who reject C. P. Snow's "Two Cultures" thesis that science and letters are intrinsically oppositional discourses. As George Levine suggests, these two discourses "support, reveal, and test each other."[5] For example, James Averill has shown the experimental and taxonomic nature of Wordsworth's language, and Hans Eichner has clarified how the Romantics supplanted Cartesian mechanical philosophy with an equally scientific view of the cosmos as organic systems.[6] Raimonda Modiano, Trevor Levere, and Neil Vickers have fruitfully explicated Coleridge's intense engagement with German science, Brunonian medicine, and English natural philosophy.[7] Answering Jonathan Smith's recent call for more studies of "specific points of contact, of actual scientists and writers," this chapter also reflects the work of Clifford Siskin, Charles Rzepka, and Brian Goldberg on Romantic professionalism.[8] Finally, Wordsworth himself invites a consideration of the contact between men of science and letters not only in his "twin labourers/same hopes" comment, but also in his declaration that a poet is, above all, "a man speaking to men," who carries "relationship" wherever he goes, and in his daring announcement that he is "ready to follow the steps of the Man of Science" should the latter ever "create any material revolution, direct or indirect, in our condition."[9]

II

Most likely, Wordsworth's rivalrous feelings toward Davy were originally provoked and then exacerbated by conversations with Coleridge and Southey, who had known Davy since he came to Bristol in 1798. For a brief, but significant time, Southey and Davy became close and ardent friends.[10] Southey engaged in Davy's nitrous oxide experiments and encouraged the young chemist to write poetry, some of which he published in his *Annual Anthologies* of 1799 and 1800. Like Coleridge and Wordsworth, Southey also used Davy as a literary advisor and editor.[11] Even in those early days, Southey's comments about his friend had a competitive edge. He told John Rickman that of all the "men of talent" in Bristol, "Davy [was] by far the first in intellect."[12] On another occasion Southey called Davy "a miraculous young man, a young chemist, a young *everything,*" and claimed, as well, that his friend "is not yet twenty-one, nor has he applied to chemistry more than eighteen months, but he has advanced with such seven-leagued strides as to over take *everybody.*"[13]

Davy's strides were, indeed, ambitiously long; he aimed at nothing less than discovering "the laws of our existence . . . [and thereby destroying] our pains and increas[ing] our pleasures," developing a "Theory of Passion," and serving not only his friends but also all of mankind.[14] When we consider how similar these utterances are to well-known passages in Wordsworth's Preface to *Lyrical Ballads* or Percy Shelley's "A Defence of Poetry," it is easier to appreciate why Romantic-era poets and scientists felt themselves to be in competition. After Davy's career-making move to London, Southey's sense of rivalry with Davy intensified, and we find him writing to a friend that "Davy, whom you know . . . as the rising pride of his county, is removed to the Royal Institute, with a good salary, where he will equally serve himself and the public."[15] A year later, as he witnessed Davy's astonishing professional progress, Southey commented archly to Rickman that "Davy of course is well and successfully employed. A new discovery of his will enrich somebody."[16]

Coleridge seems to have been almost as close to Davy in those early days in Bristol as Southey was, and he enthusiastically joined them both in Davy's notorious laughing gas experiments. John Davy believed that his brother and Coleridge were always quite close, and June Fullmer adds that Coleridge "saw Davy as a serious fellow poet."[17] Thomas Poole reports Coleridge's proclamation that had Davy not already been "the first chemist, he, probably, would have been the first poet of his age."[18] In addition to this, Joseph Cottle writes that, like Southey, Coleridge not only compared his own intellectual talents to Davy's, but he also had a distinct sense of being in competition with the energy and elasticity of Davy's mind.[19] This admiration of Davy's talent and judgment, no doubt, led the poet to call upon his friend to edit the second edition of *Lyrical Ballads* in 1800—the move that first brought Davy and Wordsworth together. During this time Coleridge noted the similarities in the genius and promise of Davy and Wordsworth, and he may have whetted Wordsworth's appetite for competition with Davy by expressing the belief that the natural philosopher was "one of the *two* human Beings of whom I dare hope with a hope, that elevates my own heart."[20] Significantly, in the early days of their careers, Coleridge thought of the efforts of both of his friends—a poet and a natural philosopher—as both similar and fundamentally ameliorative.

Not long after Coleridge made this comment, Davy delivered his sensational 1802 Discourse at the Royal Institution in London; it was a lecture that expressed many of the same values as the *Lyrical Ballads* and used many of the same linguistic and rhetorical strategies as Wordsworth and Coleridge. With this performance, Davy began to serve his age both as a spokesman and a standard-bearer for science. Although Davy's construction of his persona as the Man of Science was as much an ideal as it was a reality, the public's perception of the Davyan scientist seems to have triggered some of Wordsworth's most important claims about poetry and poets. For example, very shortly after Davy's 1802 performance, Wordsworth's second Preface to *Lyrical Ballads* appeared with several striking revisions about the figures of the Poet and the Man of Science. Conspicuous among these are his expansion and elevation of the public role of the poet as one who sings "a song in which all human beings join" and his contentious references to "the Man of Science" as an isolated worker whose knowledge is dependent on and informed by the "breath and finer spirit" of Poetry.[21] Wordsworth's rivalry with Davy may also have been exacerbated at this time by the fact that he saw how Coleridge's fascination with Davy and natural philosophy dissipated the focus and momentum Coleridge so easily lost and yet so desperately needed for his own writing and for those poetic projects he shared with Wordsworth.

From 1802 on, Wordsworth and the rest of the Romantic public were well aware of the progress of Davy's career in London and abroad. Davy's fame, Wordsworth's prior reliance upon Davy's editorial skill, and their many shared acquaintances in London and Bristol suggest why Wordsworth was willing to entertain Davy when the natural philosopher made a short tour of the Lake District in the summer of 1805, even though Wordsworth was still in deep mourning over the drowning of his brother John. Davy stayed with the Wordsworths at Dove Cottage and climbed Helvellyn with Wordsworth and another visitor, Walter Scott. This occasion is pivotal in understanding Wordsworth's relationship with Davy. John Wordsworth's death had been a great personal tragedy for his siblings, but it also affected William Wordsworth's economic and professional situation. This was because not long before John's death, William, Dorothy, and John Wordsworth had agreed among themselves that John, who was doing quite well as a ship's captain, would take on the task of earning the family's living so that William could be relieved from the financial worries that often interfered with his writing.

Reports of Davy's visit that summer vary. Dorothy Wordsworth's letters suggest that the three men had much to talk about and enjoyed their time together.[22] Mary Moorman notes that at times Davy seemed impatient or distracted to his companions and that he cut his visit short in order to hurry back to work at the Royal Institution. Based on this evidence, Moorman assumes that Wordsworth's reaction to Davy was disapproving.[23] Could not both accounts be correct? Because Wordsworth understood the similarities in their professional dreams about seeking the truths of nature and humankind, providing pleasure, and being of public service, he enjoyed conversing with the scientist. But having just had his economic plans dashed and seeing Davy so well positioned, successful, and intensely engaged in his work, it is likely that this occasion also provoked in him feelings of jealousy and professional rivalry.

A year later, Wordsworth must have been reminded once more of the contrasts between the careers of poets and scientists such as Davy. It was in 1806 that Coleridge returned from Malta in worse physical, mental, and financial shape than he had been in when he departed. Coleridge's friends, Davy and Wordsworth included, became quite concerned. The Wordsworths felt that Coleridge needed rest and medical attention. Davy tried to help Coleridge by inviting him to lecture at the Royal Institution, where, he believed, Coleridge "might be of material service to the public, and of benefit to his own mind, to say nothing of the benefit his purse might receive."[24] Turning to work and public action was characteristic of Davy when troubles arose. It is of interest that he assumed this same anodyne would work for a troubled poet. More to the point, he took it for granted that Coleridge's work and his own efforts accomplished much the same ends: public service, intellectual nourishment, and financial gain. Significantly, Coleridge considered the suggestion a reasonable one as well and excitedly wrote Davy that his lectures would cover "the most philosophical Principles . . . of Poetry."[25] Portions of this letter imply not only Coleridge's sense of professional kinship and competition with Davy, but also his belief that they addressed the same audience. Imagining his performance at Davy's place of employment, Coleridge writes: "of all men known to me, I could not justly equal any one to you, combining in one view powers of Intellect, and the steady moral exertion of them, to the production of direct and indirect Good."[26] Wordsworth was aware of Davy's lecture negotiations with Coleridge and counseled against the project. Given Wordsworth's history of professional and financial problems at this time, his awareness of Davy's advancement, and his concerns about Coleridge's instabilities, it is likely Wordsworth recognized the parallels and possibly hurtful comparisons that could be made at the Royal Institution between the work (and public successes) of Davy and the work (and public disappointments) of poets such as Coleridge and himself.

In 1806–1807 Davy's career was, in fact, progressing brilliantly. True to Southey's characterization, he was continuing to overtake everyone, poet and scientist alike. Beginning in 1795 as an apprentice to an apothecary–surgeon, by 1802 Davy had earned an appointment as professor at the Royal Institution in London. For his work on the chemistry of leather tanning he won the Royal Society's Copley Medal in 1804. This accomplishment and a series of ingenious experiments using hydrolysis earned him the honorable and very lucrative directorship of the Royal Institution's laboratory in 1805.[27] Davy's isolation of two new elements— potassium and sodium—won him international fame, including the 1807 Napoleon Prize.[28] This research and his subsequent invention of the miners' safety lamp a few years later made Davy a national hero. In 1812 came knighthood, in 1818 a baronetcy, and in 1820 his most coveted honor: the presidency of the Royal Society, a post once held by Newton.

Davy's competition with poets becomes apparent in his famous public lectures, which began in Bristol in 1799 and continued through the first two decades of the nineteenth century in London and Dublin, as well as abroad. Davy proved as compelling and passionately expressive in his scientific lectures as Wordsworth aspired to be in his poetry. In fact, Davy's discourses frequently borrowed words from the idiom of nineteenth-century philosophy, which he first encountered during his friendship with poets. Some of these words have since become associated with Romantic poetics; among them are *imagination, fancy, passion, glory, the beautiful* and *the sublime.* Favoring Wordsworth's "real language of men in a state of vivid sensation," advertised in the "Preface" to *Lyrical Ballads,* Davy not only quoted poetry, but had the added advantage of being able to illustrate his lectures with spectacular galvanic and chemical demonstrations. Davy's contemporary biographer, Dr. John Paris, reports that "[t]he enthusiastic admiration that his lectures obtained is at this period scarcely to be imagined. Men of the first rank and talent, the literary and the scientific, the practical, the theoretical, blue stockings, and women of fashion, the old, the young, all crowded, eagerly crowded the lecture-room."[29]

At no time during the early days of Davy's career did the poets of his acquaintance enjoy a fraction of the public acclaim showered on the young natural philosopher. By his own admission, Wordsworth could barely secure a "fit audience"—much less significantly lucrative returns—for some of his greatest poetic efforts, including *Lyrical Ballads* (1798–1802), *Poems in Two Volumes* (1807), *The Excursion* (1814), or *The White Doe of Rylstone* (1815). Indeed, his reviews in 1807 were so bad that during the period from 1807 to 1814, which were the years of Davy's greatest scientific achievement, Wordsworth gave up publishing poetry altogether and resorted to taking a post in the revenue service to make his living. The troubles of Coleridge's finances and writing career are well-known, as are the mixed successes and reviews of Southey's. Scott would eventually rival Davy for public acclaim (and remuneration), but not until the 1810 debut of *The Lady of the Lake.* Byron did not awake and find himself famous on a scale such as Davy's until 1812, when the first two cantos of *Childe Harold's Pilgrimage* were published.

To understand how Davy captured the public imagination as he did, we must understand that "the invention of modern science" as we know it was not yet fully underway, and that natural philosophy as most Romantics knew it was still a relatively open, amorphous, and unspecialized field.[30] This is how Davy's brother, himself an Edinburgh-educated physician, described the situation: "[natural philosophy] was just entering on that [stage] of vigorous youth; it was sufficiently advanced to display much beauty, and to excite deep interest; and it was not too much advanced to be beyond the comprehension of minds of ordinary powers devoting to it a moderate portion of time."[31] Scientists practicing in this field had much more in common with poets than they usually do today. They also seemed to understand better than we do today that both occupations are based "on man's ability to discover new combinations of experience and new methods of solving old problems."[32] Fullmer insightfully suggests that the private experiences of experimental and poetic discovery, especially as Davy and Wordsworth practiced them, have much in common:

Performing a fastidious experiment bestows an aesthetic satisfaction something like that achieved by writing a lyric poem. For a few transcendent moments a good lyric bares, fixes and illuminates an emotion. An experiment, charged with its burden of theoretical import, illuminates as if with a bright flash some part of the previously hidden natural world.[33]

Davy and Wordsworth not only recognized that they shared such theoretical and aesthetic pleasures, but they also understood their common professional needs. Not the least of these were a public forum and a receptive audience. Fullmer explains: "Poems and experiments alike, initially private in their conception, transform themselves to public acts. Without a hearer a poem is less than complete; nor is an experiment complete without an audience" (70). The necessity of going public is another key to understanding the rivalry of Romantic poets and scientists. Davy was always quite concerned about his audiences and how they might perceive science. In 1803, for example, he took pains to coach John Dalton about how to present himself as a lecturer at the Royal Institution.[34] Both the Preface to *Lyrical Ballads* and portions of the *Biographia Literaria* (1817) testify that Wordsworth and Coleridge also worried not only about audience reception, but additionally about the need to educate that audience's taste. The publics that most scientists and poets in our era address usually come from different discourse communities; such was not the case for Wordsworth and Davy. They both belonged to and competed for the attention and approval of a relatively small group of men and women who were intricately connected by family, geographical, ideological, and economic ties.[35] Moreover, this group was remarkably polymathic; as Fullmer describes them, these people "could quite casually take all of learning for [their] province . . . [and] espoused opinions on matters economic, literary, medical, political, philosophical, psychological, scientific, social and technological. What held them together were ties of friendship, of kinship, and partnership."[36]

The polymathy evident in Romantic audiences stemmed in part from educational practices of the day, which were based upon classical principles. An education of this sort inclined intellectual individuals such as Wordsworth and Davy to employ language in similar ways and to strive for the classical ideal of service.[37] Shared classical roots account as well for the fact that both men thought of themselves as philosophers. Wordsworth commonly referred to himself and Coleridge in this way and identified poetry as the "most philosophical of all writing."[38] Davy called himself a "natural philosopher," "chemical or experimental philosopher," or simply a "philosopher."[39] As self-consciously philosophical people, both Wordsworth and Davy dedicated themselves to searching for and explaining the deep and permanent truths about human nature and the natural world. They shared the conviction that their twin labors could bring pleasure and benefit to all mankind; and in this, both hoped to inherit what Shelley would later call the "civic crown."[40] Addressing their polymathic audience in new but very similar languages, using unfamiliar ideas and methods, both Wordsworth and Davy found themselves compelled to create or excite public taste for their endeavors. To this end, they chose virtually the same persona: that of the public-spirited philosopher king, or (to borrow a description Davy applied to himself in an early poem) "Poet . . . Philosopher . . . [and] . . . Sage."

Using this persona, both poets and natural philosophers of this period were forced to compete for public attention using many of the same vehicles and sites of public discourse, including the same periodical press and public lecture circuits. Most of the important periodicals of the early nineteenth century were surprisingly eclectic in their offerings of articles and reviews. To illustrate: at least one-fourth of the articles in literary journals such as *The Edinburgh Review, The Monthly Review,* and *The Quarterly Review* covered scientific subjects in depth. Early scientific journals such as *The Annals of Philosophy; or Magazine of Chemistry, Mineralogy, Mechanics, Natural History, Agriculture, and the Arts* and William Nicholson's *Journal of Natural Philosophy, Chemistry and the Arts* also published a surprising number of articles on politics, economics, history, and literature. Such a diversity of interests was usually apparent in the era's most important lecture venues as well. These included the Royal Institution, the Surrey Institute, the London Institution, the Russell Institution, and the gathering places of various Literary and Philosophical Societies, where records show that poets and scientists all spoke regularly. Because continental travel was curtailed during the war years, public lectures became a favorite pastime among educated and fashionable people. They served for a time as an important disseminating device for scientific and literary thought and helped to support many a poet or scientist. Installed prominently on the staff of the Royal Institution, Davy was especially active and well paid as a lecturer. Men of letters, who also labored in this market, albeit with varied success, included Coleridge, Southey, Sydney Smith, and William Hazlitt. No records explain why Wordsworth attended, but declined to deliver, public lectures as a young man.[41]

In sum, Davy's and Wordsworth's audiences read the same texts, attended the same lectures and social gatherings, knew both men directly or indirectly, and certainly would have recognized many of the similarities in their projects and discourse. These circumstances took on professional and economic significance for Romantic-era intellectuals because books, articles, and lectures were their bread and butter. As we know from Wordsworth's reactions to his writing blocks and negative reviews, the poet was deeply concerned about earning money as well as critical approval.[42] Competing in the literary marketplace was difficult enough. We should consider the possibility that having to compete as well with a dynamic young natural philosopher such as Davy added to the frustration and disappointment Wordsworth felt when his writing career did not go as he planned. Davy's public fame and professional heyday began in 1802 and continued for the next decade, with 1807 being one of his most glorious years. We may only speculate whether Davy's successes were a factor in Wordsworth's reaction to the treacherous reviews of his second book of poetry, *Poems in Two Volumes* (1807), which was to withdraw from publishing poetry for a number of years.[43] By the same token, it may only be a coincidence that Wordsworth's resumption of poetic publication (with *The Excursion* in 1814) occurred shortly after Davy's very public resignation from the Royal Institution as a professional natural philosopher in 1812, but the timing of these events is intriguing.

Other factors that contributed to the Wordsworth/Davy rivalry for the public role of "Poet, Philosopher, and Sage" were the real-world problems with family and finances that both men shared. Wordsworth and Davy had been born to expectations of relatively prosperous, leisurely lives. But when both boys lost their fathers as young teens, these hopes were dashed.[44] By all accounts, their fathers' deaths caused the boys great emotional and financial stress, raising issues of socioeconomic identity and eventually thrusting them into the emerging market of intellectual talent in order to support themselves.[45] The path to honorable work for young men in these circumstances was not always smooth. Wordsworth's and Davy's interests lay in fields that were evolving away from the gentleman amateur model, which was based on inherited fortunes and leisure time, to that of the occupational professional, which was based on talent, training, expertise, and payment for services rendered. Key to the new professional model for both men was the requirement of service.[46]

Additional circumstances that influenced Wordsworth's and Davy's competition to find acceptable work included increased literacy rates, the rise of periodicals and publishing houses, and the commodification of intellectual property. Jürgen Habermas's well-known discussion of the nature and transformation of the bourgeois public sphere also helps to explain the rivalry of Romantic-era poets and natural philosophers. Wordsworth, more than Davy, lamented the decline of public discourse from free and rational discussion of democratic principles to increasingly sensational and commercialized modes of expression. The poet's experience with "depraved" taste and reviewers who could make or break literary careers was particularly bitter. Nevertheless, honorable—and sometimes very lucrative—work *was* available in science and letters to young men with talent, ambition, energy, and the willingness to both educate and accommodate the reading public. Davy's letters home from Bristol and London spoke often about the opportunities for "glory" and a "genteel maintenance" that he saw opening before him. Notwithstanding his later disappointments, Wordsworth wrote his friend William Mathews in 1792 that England was a "free country, where every road is open, where talents and industry are more liberally rewarded than amongst any other nation of the universe."[47] More specifically, he told his friend that "the field of letters is very extensive" and assured him that "nothing but confidence and resolution" would be necessary for them to make their livings in London as writers.

Ironically, this professional rivalry existed before the professions of science and letters were fully in place. Both fields, however, were beginning to be discernible from each other in part because of Davy's and Wordsworth's efforts to differentiate themselves so they could compete more effectively for the same social niche and for the attention and acclaim of the same public. Sociologists believe steps such as these are characteristic of the process by which occupational groups evolve into professions. Magali Larson explains that workers in new occupations rather quickly become interest groups and, as such, become involved in the natural conflicts for place and power that exist in society.[48] Larson claims that these conflicts usually occur when workers in particular groups feel a drive for economic advantage and/or improved social status. These pressures certainly operated in Wordsworth's and Davy's lives and were made especially acute for Wordsworth as he witnessed Davy's increasing public success before their shared audience. Larson theorizes that an occupational group's drive for economic or status advantage engenders four moves that can result in that group's reconstitution as a profession: (1) alignment with power; (2) demarcation, expansion, or defense of a "jurisdiction"; (3) persuasion of the public that the group provides a valuable service and is reasonably altruistic and public spirited in its motivation; and (4) "social closure," whereby the group tries to limit access to the occupation—to its knowledge, training, credentials, market, services, and jobs—so as to protect its monopoly, to usurp the existing jurisdiction of others, and to move upward socially.[49] All of these stages can be traced in the nascent fields of Romantic science and letters and in the careers and public discourse of both Humphry Davy and William Wordsworth. This model suggests a relationship between the ways both men endeavored to differentiate themselves by redefining, expanding, and defending their jurisdictions and their service functions, and also between their respective claims to expertise. The model also supports my claim that these moves were motivated not by differences, but by functional and philosophical similarities between poets and scientists.

III

I shall now detail some of the ways in which Davy's discourse drew upon his kinship with men of letters. At the same time, as we shall see, he was beginning to mark out his professional territory and endeavoring to convince the public that the value of science and the altruism of the scientist were greater than those of literature and literary men. Two of Davy's most famous public lectures at the Royal Institution illustrate the range and timbre of his rhetoric. The first, "Discourse Introductory to a Course of Lectures on Chemistry" (January 21, 1802), launched Davy's public reputation in the learned and fashionable public sphere of London. The second, "Introductory Lecture to the Chemistry of Nature" (January 31, 1807), was delivered during one of the high points of Davy's public career, shortly after he discovered potassium and sodium.

As surprising as this may seem to us today, Davy's lectures on chemical and agricultural philosophy, galvanism, geology, and other scientific topics became important cultural and social events in London. The Royal Institution's fashionable audience caught the attention of James Gillray in 1802, at a pivotal moment in Davy's career (Figure 1.3). Although he is depicted in his previous role as assistant lecturer, Davy was named Professor of Chemistry two days before Gillray's cartoon was published, and his famous "Discourse Introductory" had appeared the previous month. As the primary lecturer, he added an element of high seriousness to the Royal Institution's program, making it even more fashionable and apparently deflecting further satire. On Davy's lecture nights, carriage traffic in the environs of the Royal Institution's Albemarle House was hopelessly snarled.[50] Inside, Davy spoke to standing-room-only crowds of four to five hundred among whom numbered not only intellectuals such as Coleridge, William Godwin, Sir George Beaumont, and the Whig reformer Henry Brougham but also some of England's most powerful men such as Sir Joseph Banks, Sir John Sinclair, Lord Spencer, Lord Somerville, the Earl of Egremont, and the Duke of Somerset. By 1804, Davy's lectures were so much in demand that the managers of the Royal Institution remodeled their building to accommodate larger audiences. Later in Davy's career, it was not unusual for enthusiasts to pay scalpers two guineas (about US$400 in today's currency) for a subscription to a course of his lectures. Another indicator of Davy's public popularity is the reaction to his near-fatal illness in the fall of 1807. The Royal Institution had to post hourly bulletins on its gates to allay the anxieties of Davy admirers, many of whom gathered each day outside Albemarle House.[51]

Davy's 1802 "Discourse Introductory to a Course of Lectures on Chemistry" caused such a stir that the Royal Institution's Managers induced Davy to publish it immediately. "No scientist since Newton," J. H. Plumb tells us, "had so captured the nation's imagination."[52] Davy was twenty-three years old. The lecture begins with a surprisingly seductive definition of chemistry as "that part of natural philosophy which relates to those intimate actions of bodies upon each other, by which their appearances are altered, and their individuality destroyed."[53] Then Davy explains how chemistry relates to or supports other areas of natural philosophy from mechanics to medicine and how it contributes to many important commercial enterprises such as agriculture and metallurgy. In all these applications of chemical philosophy, the young lecturer claims, the English citizen is served: his quality of life and work are improved, he is taught "to think and reason," and he becomes "at once the friend of nature and the friend of society" (316).

Again and again Davy uses language calculated to appeal to the public's interest in "improvement" and its concern for social tranquility, order, and permanence. Poets addressed these issues at this time as well, as we see in the early sections of Wordsworth's Preface to *Lyrical Ballads* where he worries about the great "convulsions" of their moment in history, the depravity of human judgment, and his own efforts to create poetry that will interest mankind "permanently." But Davy's stance, which was both more optimistic and less accusatory, must have been

Figure 1.3. James Gillray, *Scientific Researches*, May 23, 1802, Engraving, British Museum Catalogue No. 9923. Reproduced by permission of the Huntington Library, San Marino, CA. Davy, here depicted before his elevation to professor of chemistry, is the dark-haired young assistant holding the bellows.

more appealing. Elsewhere in the discourse, Davy implicitly engages with poets when he argues that the study or practice of chemistry is a permanent and placid "source of the most refined enjoyments and delicate pleasures of civilized society" (315). More or less friendly challenges to poets seem to be issued, as well, in his specific references to the many ways the scientist feeds man's imagination, passions, fancy, and moral constitution. Although Davy's scientist is at play in the semantic field of poets, bringing enlightenment and pleasurable experience, Davy says he also serves to "connect . . . [the] great whole of society" (323), just as Wordsworth's poet is capable of binding "the vast empire of human society" together by "passion and knowledge."[54] Davy always portrays learning about science as accessible and pleasurable. In contrast, Wordsworth advocates difficult social changes which he believes can only come in a world of readers who not only eschew any "thirst after outrageous stimulation" but who will also work long and hard to acquire the taste and talent to read the proper kind of poetry ("Preface" 599).

In the 1802 lecture Davy moves from a consideration of the general impact of science on society to a more specific "investigation of the effects of the study of this science upon particular minds," and to ascertain its powers of "increasing that happiness which arises out of the private feelings and interests of individuals" (324). Davy's statement mirrors the language and projects of not only *Lyrical Ballads* (an "experiment . . . to ascertain . . . that sort [and quantity] of pleasure a poet can provide") but also *The Prelude* (a study of the "growth of [a particular] poet's mind").[55] Because science is so various and accessible, Davy claims that it can serve not only the "man of business" and the "refined and fashionable classes of society," but also "persons of powerful minds . . . connected with society by literary, political, or moral relations" (326). Perhaps alluding to the reputation for radicalism of certain men of letters, Davy concludes his lecture by promising that scientific study gives rise to reverence, sound judgment, "tranquillity and order."

Davy's "Introductory Lecture to the Chemistry of Nature," delivered in 1807, is more tightly organized; the subject matter is more concrete; the diction is less effusive; and the speaker is clearly more confident of his reception. In short, it is more professional. The discourse begins by reminding the audience how dependent they are upon nature for their wants, comforts, and enjoyments. For this reason, Davy says, they are in need of the expertise of natural philosophers and "scientific methods" to explain the "relations and properties of natural objects, and the laws by which they are governed" (*CW* 8:169). Whereas the earlier lecture suggested that any and all of Davy's audience could practice or at least study natural philosophy, the 1807 text emphasizes the ingenuity, accumulating knowledge, and special expertise of particular men of science. He further elevates scientists above the general populace and less specialized philosophers such as poets by noting that "critical views of the general order of events taking place upon our globe" have been "corrected, enlarged, or exalted by experiments" conducted by serious, professional scientists (170).

This discourse demonstrates how far Davy has come with this professional project. Magisterially calling his subject "the great chemical economy of nature," he confidently declines the necessity of making a case for chemistry, nor does he feel the need to dress his "homely" subject up with "sumptuous, elegant, amusing," or "wit-exciting" language (179).[56] In what is, most likely, a response to charges that scientists are solitary beings shut away in forbidding laboratories, as Wordsworth implies in the Preface to *Lyrical Ballads* or as Mary Shelley will later articulate in *Frankenstein*, Davy invokes the authority of Bacon and Newton, England's most revered natural philosophers, who were also known as men of letters, to defend science. Then he insists that chemical philosophy is not just "laborious operations in crucibles" or "[operations] of artificial processes in phials, retorts, or alembics" (179). Instead, it is a study of

> the great forms and elements of the external world, which are at once objects of vulgar admiration, of imitation in art, and of poetical description. Sunshine, winds, vapours, clouds, rivers, and cataracts are its prime agents; and the scenes of their operations, the diversified face of nature, the sky, the ocean, mountains, plains, and valleys. (180)

Such language clearly asserts the scientists' jurisdiction not only over the study of nature but also over the very images and representations from which Romantic poets crafted some of their greatest poems.

Davy sanguinely compares the "greatness of modern times," which resides in "the state of our physical knowledge, and the chemical and mechanical inventions and arts dependent upon its principles" (181), to the glory of Greece and Rome with their heroes and conquerors, poets, orators, and artists. He recalls the decadent phase of classical literature and art, a gesture that may be taken as an implicit warning to his contemporaries in these fields: "Literature and . . . fine arts arrived at a degree of perfection beyond which even ambition had nothing to desire; and when the turbulence of war had ceased, when the restlessness of conquest had passed away, they became only as roses, strewing the path that led to luxury and ruin" (181). In contrast, he claims that Romantic era science has been able, through the "processes and art of experiment," to preserve the "mind in a continual state of activity" and constantly to produce "new fields for investigation" (181–82). As a result, in his day "literature has been an instrument of science; science has given new ideas and new combinations to literature; and even the objects of the imitative arts have been extended in consequence of experimental research" (182). This assertion of scientific priority over the arts is followed by an equally proprietary suggestion that only science can preserve "the equilibrium between reason and feeling" (182). Davy continues an implicit critique of contemporary men of letters such as Wordsworth (who, at times, seemed quite vulnerable to the charge of worshiping the idols of his own imagination) and Coleridge (whose great creative talent seemed to visit and disappear), saying:

> Men of science, instead of worshipping idols existing in their own imaginations, have examined with reverence and awe the substantial majesty of nature. Discovery has not visited them and disappeared again, like the flashes of lightning amidst the darkness of night; but it has slowly and quietly advanced, as the mild lustre of the morning promising a glorious day. (182)

The thrumming rhythm and culminating image of promise and glory in Davy's conclusion are characteristic of his more mature discourse, and they testify to how effectively this scientist could mobilize the tools of the poet to promote and valorize science.

I should like to point out that competition between men of science and letters is a constant in both of Davy's lectures. This is no accident, but rather a function of the early kinship of Romantic science and letters and this polymathic moment in British history. Second, the tone of the contest changes in the five intervening years between these two discourses. In 1802 Davy is friendly and fairly collegial toward men of letters. By 1807 he has become more confident, but also scrappily territorial. This may be attributed to quirks of Davy's personality and his precipitous ride to personal celebrity; but according to Larson's theory, this is also a sign of significant progress in the project of professionalization. The professionalization of science was, of course, driven in part by the proliferation of knowledge; but it was also necessitated in part by the rivalry between poets and scientists for the civic crown. The stages seem clear: science's alignment with power was already in place at the Royal Institution long before 1807; from 1802 to 1807 Davy carved out a jurisdiction and aggressively marketed the many philanthropic services of chemistry and the other branches of scientific knowledge. In addition to all this, he became increasingly secure socially and financially.

At this same time, Wordsworth's progress toward professionalizing poetry was faltering. No equivalent of the Royal Institution was in place as a platform for poets. No board of interested gentlemen implored Wordsworth to rush his creative products—which included some of his most enduring lyrics and the 1805 *Prelude*—into publication. Unlike Davy, behind whom stood numerous individuals who were likely to profit materially from his intellectual labors, Wordsworth had no riches to offer the privileged class nor clearly marketable skills to teach the artisan class. Indeed, Wordsworth was not yet perceived—nor did he think of himself—as a professional success. He held his finest poem, *The Prelude,* back from publication, believing that the story of his life would not be of interest until his great work, the imagined but never fully realized *The Recluse,* had made him famous. The poems Wordsworth did elect to publish in 1807, *Poems in Two Volumes,* were denounced by no less than eighteen different reviewers.[57] Moreover, his financial and social situation was still precarious. Despite the Lonsdale settlement in 1802, Wordsworth had invested the money unwisely, so he was again pressed for money to feed his growing family. He found it necessary to petition friends to make good on investments he had made with them and welcomed the Beaumonts' offer to put his family up at Coleorton during the winter of 1806–1807.

If we accept the notion that Wordsworth and Davy, as standard-bearers of letters and science, were locked in a contest for the philosopher king jurisdiction, then the sociological model of professionalization sheds light on what the poet's problem was. He had no alliance with power as Davy had (indeed, he felt it was his duty at times to critique that power); nor could he specify or regulate professional standards in the same way Davy could. He was unable to control or predict the economic returns for his professional efforts; and when remuneration did come for Wordsworth's early literary work, it was usually meager. Years later, Moorman notes that Wordsworth listed three reasons for his failure to provide for his family as a professional poet: "'the pressure of the time,' when prices were rising; his own miscalculation of 'the degree to which my writings were likely to suit the taste of the time'; and lastly the fact that his most important works would not be ready for publication for many years."[58] In contrast, the pressures of the times, rising prices, and the taste of the era all worked in Davy's favor; moreover, Davy's professional efforts quickly produced marketable ideas.

An additional point about Davy's professionalizing moves is noteworthy. To outmaneuver the poets for jurisdiction in the area of public service, Davy shrewdly played on the anxieties of the English people. British angst at this time was, of course, occasioned by hostilities with France and all the hardships that came with war. Pressing this particular hot button was an audacious move, for Davy and other English scientists were known to communicate—through letters, journal publications, and so forth—with their counterparts on the Continent, especially with French researchers such as Laplace, Berthollet, Carnot, or Monge, who were variously involved with Napoleon's government and army.[59] Davy could have been a target of reactionary criticism for seeming to sympathize with the French, as so many of the literati, and certain unfortunate scientists such as Joseph Priestley and Thomas Beddoes, had been. Davy took care to deflect reactionary attack, however. He always spoke of English natural philosophy as a national resource and linked "the glory of our science" with the "glory of our own country."[60] The caginess of Davy's other rhetorical strategies to outflank reactionary criticism and gain precedence over literary people is evident in his many compliments to his auditors. He often told them that their interest in natural philosophy meant they were more civilized, rational, peaceful, and God-fearing people than other Europeans. He assured them that they would ride out the "great changes," "convulsions," and "caprice" of the day more easily because of their knowledge of science (*CW* 2:311). With such an audience, these appeals were likely to have been more rhetorically effective than Wordsworth's glum condemnation of their "degrading thirst after outrageous stimulation" or his assertion that he was more sensitive, tender, and knowledgeable about human nature than they ("Preface" 599).

So effective was Davy at protecting English science and scientists from being associated with French radicalism that he was not seriously criticized for his journey to Paris in 1807—at the height of the war with France—to collect the Napoleon Prize for his discovery of potassium and sodium. This is how Davy's contemporary biographer, John Paris, described the effect upon England of Davy's winning the Napoleon Prize:

> Let the reader only recall to his recollection the bitter animosity which France and England mutually entertained towards each other in the year 1807, and he will be able to form some idea of the astounding impression which the Bakerian Lecture must have produced on the Savans of Paris, when, in despite of national prejudice and national vanity, it was crowned by the Institute of France with the prize of the First Consul! Thus did the Voltaic battery, in the hands of the English chemist, achieve what all the artillery of Britain could never have produced—a SPONTANEOUS AND WILLING HOMAGE TO BRITISH SUPERIORITY![61]

Although Golinski has rightly commented that Davy's success was due in part to his conservative political alliances and willingness to be a mouthpiece for the Royal Institution board, Davy's public discourses demonstrate very real concern about social injustice and class disparities at home. They also make it clear that expressions of this sort, for which Romantic writers often drew fire, could be successful with British audiences of the early nineteenth century. In his public response to the domestic situation, Davy seems to have understood the contemporary concerns of both the working and the landed classes. Along with this, he mixes sympathy for the fears of the establishment with a qualified appreciation of aspects of the republican politics once espoused by intellectuals and friends such as Beddoes, Coleridge, and Wordsworth. To do this Davy offers chemistry as a means of social relief for the poor and introduces the man of science as mediator and teacher of the rich. His goal is to foster "a state of society in which the different orders and classes of men will contribute more effectively to the support of each other than they have hitherto done" (*CW* 2:322).

An additional contribution to Davy's success may have been his astute analysis of the public he and Wordsworth shared. Writing a friend in Cornwall in January 1801, Davy notes: "Never was the state of public affairs in England more confused than at this moment, and never were hopes of peace and plenty feebler in the public mind" (1:107). Recognizing an opportunity, Davy offered science as a new arena for positive and fruitful endeavor. Calling on the Enlightenment ideal of progress and using a typically Romantic figure of society as a living organism, he says,

Science has done much for man, but it is capable of doing still more; its sources of improvement are not yet exhausted; the benefits that it has conferred ought to excite our hopes of its capability of conferring new benefits; and in considering the progressiveness of our nature, we may reasonably look forward to a state of greater cultivation and happiness than that we at present enjoy. (2:319)

In addition to all of this, Davy made the quest for scientific knowledge seem both excitingly new and reassuringly familiar: "The future is composed merely of images of the past, connected in new arrangements by analogy, and modified by the circumstances and feelings of the moment; our hopes are founded upon our experience" (2:320).

Consider how much more appealing rhetoric such as this must have been to the period's audience when compared to some of Wordsworth's criticisms of English society in poems about vagrants, ruined cottages, discharged soldiers, and the last of a poor shepherd's flock. Indeed, part of Davy's appeal was that he offered many of the same gains promised by Romantic poets without the pain. Like the poets, Davy described the sublime and beautiful operations of nature; he also experimented with individual sensation, offered sensitive observations of the natural world, marshaled the imagination and the intellect for the service of society, promised pleasure, and traced the organic connectedness and affinities of all life. In addition to all this, however, Davy offered hope for tangible improvement in the quality of daily life for his countrymen and very real opportunities for people to make money. He ousted the pain usually associated with gain by representing scientific improvement as familiar, exciting, and requiring no real change in human behavior or English social structures. Counting on the human love of learning, he promised that the process of bringing about a new world would be congenial, natural, and gentle: "the influence of true philosophy will never be despised; the germs of improvement are sown in minds even where they are not perceived, and sooner or later the springtime of their growth must arrive" (2:322).

In these ways Davy was able to outmaneuver his less strategic literary rivals for the interest, acclaim, and loyalty of their shared audience. Beyond this, Davy had two powerful tools unavailable to most of the poets: the site and paraphernalia of his discourses. The Royal Institution's beautiful Albemarle House possessed a magisterial lecture hall with some of the best acoustics in London. Davy addressed his audience from an altarlike table on a raised dais in that hall. On the table he displayed the instruments of science—retorts and crucibles, scales and pneumatic troughs—and reproduced spectacular electrochemical experiments. Behind him the entryway to the laboratory itself revealed his great furnaces, shelves of chemical vials, and huge voltaic piles (Figures 1.4 and 1.5, cp. 1.3). Harriet Martineau remarked that when Davy lectured, he "presented most strongly to the popular observation the attributes of genius."[62] Golinski has added that Davy's use of this setting and its equipment not only displayed his genius and the successes that

Interior View of the Laboratory in the Royal Institution.

Figure 1.4. W. T. Brande, *A Manual of Chemistry* 2nd ed. (London: John Murray, 1830), Plate II: Interior View of the Laboratory in the Royal Institution. Engraving by James Basire from a drawing by William Tite. Courtesy of the Linda Hall Library, Kansas City, MO.

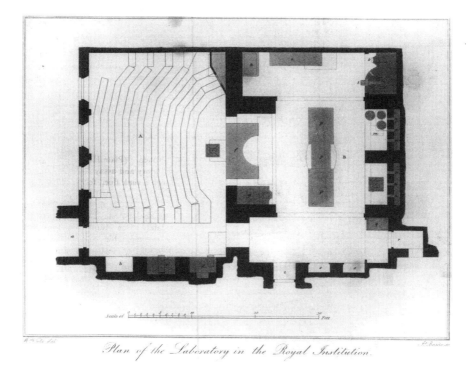

Figure 1.5. W. T. Brande, *A Manual of Chemistry* (London: John Murray, 1819), Plate I: Plan of the Laboratory in the Royal Institution. Engraving. Reproduced by permission of the Syndics of Cambridge University Library.

chemistry had achieved, but it also "established acceptance among his audience that the powers of nature were . . . Davy's to command."[63] This was not only scientific professionalism; it was a species of theater or performance art—small wonder that Davy captured not only the Romantic audiences' intellectual attention but also their emotional and aesthetic interests in ways that few poets ever did.

IV

The purpose of this account of Davy's rivalry with Wordsworth has been twofold: to enrich scholarly understanding of the scientific contexts of Romantic literary production and to invite scholars to consider these contexts when they evaluate the terms of the Romantic poets' construction of themselves and their profession ("relations [always social] determine terms"). For example, Wordsworth's relationship with Davy explains the genesis of this perplexing definition of genius in the fine arts, which appears in Wordsworth's *Essay, Supplementary to the Preface* (1815):

> Of genius, in the fine arts, the only infallible sign is the widening [of] the sphere of human sensibility, for the delight, honor, and benefit of human nature. Genius is the introduction of a new element into the intellectual universe: or, if that be not allowed, it is the application of powers to objects on which they had not before been exercised, or the employment of them in such a manner as to produce effects hitherto unknown. What is all this but an advance, or a conquest, made by the soul of the Poet?[64]

The three metaphors employed above bear an uncanny similarity to the story of Davy's rise to fame: his discovery of new elements, his inspired and unprecedented application of the power of Volta's battery to chemical compounds, and the dramatic effects he produced by replicating his galvanic experiments in his lectures. Wordsworth's odd transformation of these figures into a conquest of the Poet's soul makes more sense if we grant that his relationship with scientists was both sibling-like and rivalrous.

Besides this passage, the language and dramatic situation in some of Wordsworth's sonnets, longer lyrics, and portions of *The Prelude* echo his rivalrous relation to Davy. For example, the third book of *The Prelude* is especially ripe for rereading in light of the Wordsworth–Davy rivalry. Might Wordsworth's references to the "excessive hope,/Tremblings withal and commendable fears,/Small jealousies and triumphs good or bad" of Cambridge students refer to the shared—not opposite—hopes of natural and poetical philosophers?[65] Are the jealousies small because their differences are small? Moreover, when the poet notes that the statue of Newton, "with his prism and silent face" in the Trinity College chapel opposite his St. John's room, was a daily presence for him, we can now appreciate that this may not be merely a description of a particular setting, but that it may also be a powerful reminder of the complicated professional challenges and emerging rivalries with which this young poet, and indeed all other people of letters *and* science at Cambridge, lived. A reading such as this seems likely, considering not only that Wordsworth left Hawkshead School (a school distinguished by the excellence of its instruction in mathematics and natural philosophy) with the expectation of excelling in mathematics, but also that St. John's and Trinity Colleges were considered to be the strongholds of mathematics and science at Cambridge.

As a final example of how illuminating an appreciation of the sibling-like rivalry between Wordsworth and Davy can be in studies of Romantic texts, let us consider the startling conclusion of the 1805 *Prelude*. Surely the language and images of Humphry Davy's personal and professional successes loom behind

Wordsworth's resounding announcement of what, finally, he has learned about himself and his work from writing *The Prelude*. He retools the word "labourer"—now he and Coleridge alone are described as "joint labourers" and "Prophets of Nature" (13:443, 13:446). He also declares that they will "speak/A lasting inspiration," which is "sanctified" as much as any of Davy's inspiring scientific performances "By reason and by truth" (13:446–48). They will teach their fellows that the "mind of man," a province he claims for the poet, remains "unchanged" "'mid all revolutions in the hopes/And fears of men" (13:453–54). This mind—this creative, *poetical* mind—is not only a "thousand times more beautiful than the earth," a province now surrendered to the scientist, it is also "*of substance . . . more divine*."[66]

Given the similarities between early Romantic scientists and poets, the rivalry these similarities caused, the nature of the professionalizing process as sociologists describe it, and Davy's brilliant public career, it is no surprise that Wordsworth sometimes felt overwhelmed and faltered in his writing career. These factors help to explain why he eventually defined the poet not as a professional man, nor even as a lover of nature, but as a purely intellectual and spiritual being (see Figure 1.2 in which the portraitist highlights these qualities). Indeed, if Larson's analysis holds, then technically a true *profession* of poetry—especially in comparison to the emerging profession of science—did not exist in the early Romantic period.[67] Finally, considering the nature of Wordsworth's idealism about the kinds of public service he aspired to render, the difficulty of predicting or dictating to people what is pleasurable, and contemporaneous reforms in education that opened reading and writing to an ever-widening circle, Wordsworth could not, nor would he have wanted to, professionalize the literary life as Davy was endeavoring to professionalize the scientific one.

Notes

The epigraph that opens this chapter is from *The Practice of Everyday Life* (Berkeley: U of California P, 1984), ix.

1. Wilfred S. Dowden, ed., *The Journal of Thomas Moore* (Newark: U of Delaware P, 1983), 355–56.

2. Jonathan Wordsworth, M. H. Abrams, and Stephen Gill, eds., *William Wordsworth: The Prelude 1799, 1805, 1850* (New York: Norton, 1979), Book Fifth, lines 41–43.

3. See "Preface" to *Lyrical Ballads* in Stephen Gill, ed., *The Oxford Authors: William Wordsworth* (Oxford: Oxford UP, 1990), 606.

4. See Jan Golinski, *Science as Public Culture: Chemistry and Enlightenment in Britain, 1760–1820* (Cambridge: Cambridge UP, 1992); and David Knight, *Humphry Davy: Science and Power* (1992; Cambridge: Cambridge UP, 1998).

5. George Levine, "The Novel as Scientific Discourse: The Example of Joseph Conrad," *Novel* 21 (1988): 223.

6. James Averill, "Wordsworth and 'Natural Science': The Poetry of 1798," *Journal of English and Germanic Philology* 77.2 (1978): 232–46; Hans Eichner, "The Rise of Modern Science and the Genesis of Romanticism," *PMLA* 97 (1982): 18–30.

7. Raimonda Modiano, *Coleridge and the Concept of Nature* (Tallahassee: Florida State UP, 1985); Neil Vickers, "Coleridge, Thomas Beddoes and Brunonian Medicine," *European Romantic Review* 8.1 (1997): 47–94. Trevor H. Levere's many studies of Coleridge and science include *Poetry Realized in Nature: Samuel Taylor Coleridge and Early Nineteenth-Century Science* (Cambridge: Cambridge UP, 1981), as well as several essays.

8. See Jonathan Smith, *Fact and Feeling: Baconian Science and the Nineteenth-Century Literary Imagination* (Madison: U of Wisconsin P, 1994), 5–6; Clifford Siskin, "Wordsworth's Prescriptions: Romanticism and Professional Power," *The Romantics and Us*, ed. Gene Ruoff (New Brunswick: Rutgers UP, 1990), 303–21; Charles Rzepka, "'A Gift that Complicates Employ': Poetry and Poverty in 'Resolution and Independence,'" *Studies in Romanticism* 28.2 (1989): 225–47; and Brian Goldberg, "'Ministry More Palpable' William Wordsworth and the Making of Romantic Professionalism," *Studies in Romanticism* 36.3 (1997): 327–47.

9. "Preface" to *Lyrical Ballads* in Gill, *Oxford Authors: Wordsworth*, 603, 606–07.

10. Southey's enthusiasm about Davy moved him to write Coleridge in 1799 that Davy "is one of my deities" (Kenneth Curry, *New Letters of Robert Southey* [New York: Columbia UP, 1965], 209). For more on Davy and the young poets of Bristol, see June Fullmer, *The Young Humphry Davy: The Making of an Experimental Chemist* (Philadelphia: American Philosophical Society, 2000), especially 121–34.

11. In 1800, Southey entrusted *Thalaba* to Davy for revision (Curry, *New Letters*, 234).

12. Charles Cuthbert Southey, *The Life and Correspondence of the Late Robert Southey*, 2 vols. (London: 1850), 2:45–46.

13. Ibid., 2:13.

14. John Davy, ed., *The Collected Works of Sir Humphry Davy*, 9 vols. (London: Smith, Elder, 1839–1840), 1:77, 91, 57.

15. Curry, *New Letters*, 247.

16. Ibid., 269.

17. *The Young Humphry Davy*, 137.

18. Quoted in John Davy, *Fragmentary Remains, Literary and Scientific, of Sir Humphry Davy* (London: Churchill, 1858), 322–23.

19. Joseph Cottle, *Reminiscences of Samuel Taylor Coleridge and Robert Southey* (Highgate: Lime Tree Bower, 1847), 329.

20. Earl Leslie Griggs, ed. *Collected Letters of Samuel Taylor Coleridge*, 3 vols. (Oxford: Clarendon, 1956), 1:773–74; emphasis added.

21. Gill, *Oxford Authors: Wordsworth*, 606. See also Roger Sharrock's discussion of Davy's 1802 discourse and Wordsworth's prefaces in "The Chemist and the Poet: Sir Humphry Davy and the Preface to the *Lyrical Ballads*," *Notes and Records of the Royal Society* 16 (1961): 57–76.

22. See Dorothy's letter to Lady Beaumont (August 26, 1805) in Ernest De Selincourt, ed., *The Letters of William and Dorothy Wordsworth, Volume 1: The Early Years, 1787–1805*, 2nd ed. (Oxford: Clarendon, 1967), 620–21.

23. Mary Moorman, *William Wordsworth, a Biography*, 2 vols. (Oxford: Clarendon, 1957), 1: 57–58.

24. Davy, *Fragmentary Remains*, 98.

25. In fact, the lectures he planned were impossibly ambitious, covering writers from Chaucer to Thomas Chatterton, and "all I know . . . on the subjects of Taste, Imagination, Fancy, Passion, the source of our pleasures in the fine Arts . . . and the connection of such pleasures with moral excellence." See Griggs, *Collected Letters,* 2:30.

26. Ibid., 2:28.

27. Besides room and board, Davy was paid £400 a year—more than twice the £150 annuity the Wedgwoods gave Coleridge in 1798. It was not until 1813—eight years later—that Wordsworth began to earn anything like this sum for laboring, not as a poet, but as the Distributor of Stamps for Westmoreland.

28. Not only did the prize bring international fame of the highest order, it also carried with it an award of 3,000 francs (approximately £120).

29. "Humphry Davy," *Dictionary of National Biography,* 63 vols. (Oxford: Oxford UP, 1981), 5:639.

30. The "invention of modern science" is Andrew Cunningham and Perry Williams's term for what has also been called the "Second Scientific Revolution," which is usually dated from the early Romantic period and extends into the late Victorian period. See Andrew Cunningham and Perry Williams, "De-centring the 'big picture': *The Origins of Modern Science* and the modern origins of science," *British Journal for the History of Science* 26 (1993): 407–32. Cunningham and Williams define "modern science" as a new, secular discipline for studying natural phenomena that is characterized by free inquiry, meritocracy, material progress, and specialization. Part of what made it modern was the support of new technologies, research institutions, government, industry, and universities.

31. John Davy, *Memoirs of the Life of Sir Humphry Davy, Bart.*, 2 vols. (London: Longman, Rees, Orme, Brown, Green, and Longman, 1836), 1:154.

32. Elliot A. Krause, *The Sociology of Occupations* (Boston: Little, Brown, 1971), 256.

33. *The Young Humphry Davy*, 69–70.

34. Henry Bence-Jones, *The Royal Institution, Its Founder and Its First Professors* (London: Longman, 1871), 334.

35. Edmund Burke estimated that in the last decade of the eighteenth century, England had no more than eighty thousand readers. School and university enrollment figures for the period support this notion of the public's small size: on average only 5 percent of all children remained in school after age eleven; in 1801 the combined student bodies of Oxford and Cambridge numbered a little more than twelve hundred undergraduates. See John Lawson and Harold Silver, *A Social History of Education in England* (London: Methuen, 1973); and L. S. Sutherland and L. G. Mitchell, eds., *The History of the University of Oxford, Volume V: The Eighteenth Century* (Oxford: Clarendon, 1986).

36. Fullmer, *The Young Humphry Davy,* 122. See also my "Rivals in the Public Sphere: Humphry Davy and Romantic Poets," diss., U of Texas at Austin, 1998, 106–11 (on networks of acquaintance in the Romantic public sphere) and 187–218 (on polymathy).

37. Wordsworth's schooling at both Hawkshead Grammar School and St. John's, Cambridge, was solidly classical. While Davy did not attend university as Wordsworth

did, notes from his apprenticeship show that he was tutored extensively in classical subjects such as ethics and "moral virtues," logic, rhetoric and oratory, classical languages (Latin, Greek, and Hebrew), mathematics, physics, and mechanics. Moreover, his boyhood mentors, Dr. John Tonkin and his brother the Reverend William Tonkin, were classically educated and worked hard to instill in their surrogate son the values of this system of thought. See Fullmer (*The Young Humphry Davy*, 9–74) and Knight (*Humphry Davy*, 1–26) on Davy's education.

38. "Preface" to *Lyrical Ballads* in Gill, *Oxford Authors: Wordsworth*, 605.

39. In fact, the word *scientist* as it is known today was not yet in use. See Sidney Ross, "The Story of a Word," *Nineteenth-Century Attitudes: Men of Science*. Vol. 13: *Chemists and Chemistry* (Dordrecht: Kluwer Academic, 1991), 1–39.

40. See "A Defence of Poetry," in Donald H. Reiman, ed., *Shelley's Poetry and Prose, Authoritative Texts, Criticism* (New York: Norton, 1977), 500, where Shelley notes that "the poets have been challenged to resign the civic crown to reasoners and mechanists."

41. For more on public lecturing, see J. N. Hays, "The London lecturing empire, 1800–50," *Metropolis and Province: Science in British Culture, 1780–1850*, ed. Ian Inkster and Jack Morrell (London: Hutchinson, 1983), 91–119; Larry Stewart, *The Rise of Public Science: Rhetoric, Technology, and Natural Philosophy in Newtonian Britain, 1660–1750* (Cambridge: Cambridge UP, 1992); and Thomas A. Kelly, *A History of Adult Education in Great Britain* (Liverpool: Liverpool UP, 1992).

42. Both Dorothy and John Wordsworth believed that a number of William's physical ailments were attributable to anxiety over his work and his ability to make a living as a poet. We might judge his attitude about writing for money by his reaction to Southey's less-than-favorable review of *Lyrical Ballads*. He expressed his irritation with Southey over the review in a letter, saying Southey knew that "I published these poems for money and money alone. He knew that money was of importance to me." In the same letter, Wordsworth also commented, "I care little for the praise of any other professional critic but as it may help me to pudding" (Moorman, *Wordsworth*, 1:442).

43. From 1808 until 1810, Wordsworth published only prose: *The Convention of Cintra* (1808–1809), *Reply to Mathetes* (1809–1810), and *Guide to the Lakes* (1810).

44. Davy's father, who has been described as a "yeoman," married into a family of prosperous merchants who possessed freehold property. He died in 1794, when Davy was in his teens, ending the boy's dreams and an Edinburgh University education and thrusting him almost immediately into an apprenticeship. Wordsworth's financial circumstances and family drama are well-known. See especially Moorman, *Wordsworth*; Stephen Gill, *Wordsworth, A Life* (Oxford, Oxford UP, 1990); and Kenneth Johnson, *The Hidden Wordsworth: Poet, Lover, Rebel, Spy* (New York: Norton, 1998).

45. See Goldberg ("Ministry More Palpable") on Wordsworth's identity issues vis-à-vis his profession. See John Paris, *The Life of Humphry Davy* (London: Colburn and Bentley, 1831); Fullmer (*The Young Humphry Davy*); and Golinski (*Science as Public Culture*) for Davy's issues with family and identity.

46. See Morris Berman, *Social Change and Scientific Organization: The Royal Institution, 1899–1844* (London: Heineman, 1978) on gentlemen amateurs of science. See Philip Elliot, *The Sociology of the Professions* (London: Macmillan, 1973) on occupational profes-

sionals. For more on science, the gentleman, service, professional fees, and status in the early nineteenth century see Daniel Duman, "The Creation and Diffusion of a Professional Ideology in Nineteenth Century England," *The Sociological Review* 27.1, new series (1979): 113–38; and Joseph Ben-David, *The Scientist's Role in Society: A Comparative Study* (Englewood Cliffs, NJ: Prentice-Hall, 1971) and "The Profession of Science and Its Powers," *Minerva* 10.3 (1972): 362–83.

47. Ernest de Selincourt, *The Letters of William and Dorothy Wordsworth, the Early Years 1787–1805*, 2nd ed. (Oxford: Clarendon, 1967), 75.

48. See Magali S. Larson, *The Rise of Professionalism: A Sociological Analysis* (London: U of California P, 1977). While Larson's work on the professions is considered definitive, Keith Macdonald has recently updated Larson's theories. I draw my examples from Macdonald's *The Sociology of the Professions* (London: Sage, 1995).

49. Macdonald, *The Sociology of the Professions*, 28–29.

50. Founded in 1799 by Count Rumford with several prominent Englishmen, the Royal Institution was the first freestanding scientific research institute of its kind in Britain. Located in Mayfair, supported by generous subscriptions of England's wealthiest families, the Royal Institution attracted a very fashionable and literary audience for its public lecture series. See Bence-Jones (*The Royal Institution*), Berman (*Social Change and Scientific Organization*), and Golinski (*Science as Public Culture*). For details relevant to the cartoon, see M. D. George, *Catalogue of Political and Personal Satires Preserved in the Department of Prints and Drawings in the British Museum*, 11 vols. (London: British Museum, 1978), 8:112–14 (# 9923).

51. Letters Dorothy Wordsworth wrote during the winter of Davy's illness testify to her brother's awareness of and concern about Davy. See letters 90, 91, and 93 (December 2, 1807–January 3, 1808) in de Selincourt, *Wordsworth Letters, Volume III: Middle Years, Part 2*.

52. See J. H. Plumb, *England in the Eighteenth Century* (Harmondsworth: Penguin, 1950), 169.

53. Davy, *Collected Works*, 2:311. Cited parenthetically in the text hereafter as *CW*.

54. "Preface" to *Lyrical Ballads*, in Gill, *Oxford Authors: Wordsworth*, 606.

55. *The Prelude* was not published in Wordsworth's lifetime, but it nevertheless occupied the thinking and work of both Wordsworth and Coleridge throughout their careers. Moreover, it is likely that Davy knew about the text, for he visited Wordsworth in the summer of 1805 when Wordsworth was completing the second version of the poem.

56. Similarly, Wordsworth had claimed in the Preface to *Lyrical Ballads* that poets found it unnecessary to "trick out" natural things.

57. See Katherine Mary Peek, *Wordsworth in England: Studies in the History of His Fame* (New York: Octagon, 1969), 30–48.

58. Moorman, *Wordsworth*, 2:243.

59. See my "Rivals," 29–33.

60. The Royal Institution's board of proprietors and managers had specifically requested restraint from its employees on political matters. In a letter to his friend Davies Giddy, written in March 1801, Davy notes that "Count Rumford professes that [my position at the Royal Institution] will be kept distinct from party politics. I sincerely wish that such may be the case, though I fear it" (Bence-Jones, *The Royal Institution*, 320–21).

61. *The Life of Humphry Davy*, 165.

62. Harriet Martineau, *A History of the Thirty Years' Peace*, 2 vols. (London, 1877), 594.

63. *Science as Public Culture*, 200.

64. Gill, *Oxford Authors: Wordsworth*, 269–70.

65. Wordsworth et al., *The Prelude*, Book Third, lines 66–68; emphasis added.

66. Line 13:445; emphasis added. Davy died in 1829; in the 1850 version of *The Prelude*, Wordsworth changed "substance" to "quality."

67. Though there certainly were professional writers at this time who wrote poetry, such as Charlotte Smith and Anna Barbauld, their income came chiefly from prose.

CHAPTER TWO

The Rock Record and Romantic Narratives of the Earth

NOAH HERINGMAN

> We must read the transactions of times past in the present state of natural bodies.
>
> —James Hutton, *Theory of the Earth* (1795)

The bodies in question in Hutton's *Theory of the Earth* are rocks, which resist reading, but transact a great volume of business. The rocky landforms of Romantic poetry—Mont Blanc, the Simplon Pass, Ben Nevis—also famously resist reading, generating images that articulate the otherness of the physical through the literal and metaphorical opacity of rock. This aesthetic response to the materiality of rocks and landforms is, however, inseparable from the emerging economic category of natural resources. Hutton's criterion of legibility is developed further by the more empirical geologists of the following generation, especially William Smith. Smith's economic geology preserves the paradox of a landscape both profitable and "romantic," a paradox expressed most vividly in the range of Percy Shelley's descriptions. While Shelley describes Mont Blanc as an alien, "unearthly" landscape, the earth itself is domesticated as an "infinite mine" in his *Prometheus Unbound* (1820). The model of an "infinite mine," with its latent natural history, generates what might be called a historiography of the earth.

John Clerk's famous engraving from Hutton's *Theory* gives a visual form to this historiography (Figure 2.1). As a heuristic representation of the rock record, it distinguishes three basic phases of earth history. The engraving shows the side of a cliff exposing a set of vertically inclined strata underlying more recent, horizontal strata. If we understand these lower strata as having been uplifted and tilted from their original horizontal position, the image conveys some idea of the tremendous degree of

Figure 2.1. James Hutton, *Theory of the Earth*, 2 vols. (London: Cadell and Davies, 1795), Vol. I, Pl. III: Jedburgh Unconformity. Engraving by D. B. Pynt from a drawing by John Clerk of Eldin. Courtesy of the Linda Hall Library, Kansas City, MO.

subterranean heat and pressure by which, as Hutton argues, rocks are transformed and continents elevated. The deformation of these lower strata, especially to the right, evokes these immense plutonic forces, while the delicate and miniaturized landscape above suggests the contrasting "imbecility" of human capacities, in Hutton's phrase. At the same time, this is recognizably the landscape of "improvement," and the image seems to figure the history of the earth as a narrative of improvement. William Smith was able to read more definitely, and to capitalize more fully on geological transactions, by focusing on the orderly sequence of horizontal strata. Through his system of fossil "characters," Smith claimed, the strata became "more intelligible and useful." When the history the earth has recorded in itself becomes legible, it is found to be a history of improvement, culminating in the human transformation of landscape. The tremendous age of the earth, of all the features of the earth's material so widely discussed during the Romantic period, becomes the province of an increasingly scientific geology, and the "rock record" today is still the basic paradigm of historical geology. In its early-nineteenth-century form—particularly in the fossil archives of Smith and Shelley—the rock record represents the earth simultaneously as the substance and the text of history, generating a materiality located precisely between the two materialities recently competing for the objects of Romanticism, that of the letter and that of history.[1]

Toward the end of *Prometheus Unbound*, an apocalyptic light illuminates "the secrets of the Earth's deep heart,/infinite mine of adamant and gold," a fund of wealth embedded in a subterranean archive of prehistoric fauna and cultures (4.279–80). The numerous scientific allusions in this passage have been traced to James Parkinson and Humphry Davy.[2] Davy's popular lectures on geology, given repeatedly between 1805 and 1811, provide a representative description of the rock record for this period. The "secondary strata" are, according to Davy,

> monuments of the great changes the globe has undergone. They exhibit indubitable evidences of a former order of things and of a great destruction and renovation of living beings. . . . The connection between their causes and effects is obscure but apparently within the reach of our faculties, and it is displayed in characters which can be deciphered only with difficulty, but which express sublime truths.[3]

Shelley's light brings these phenomena *entirely* within reach of "our faculties," augmenting their sublimity at the same time. Similarly, Smith's famous map and his prose (1815–1817) maintain the aesthetic provenance of the rock record while offering decipherable "characters" in the strata's fossils. Despite their differing idioms, both these accounts expand the scope and utility of a rapidly accumulating body of geological knowledge, linking it to existing notions of social and economic progress. Both Smith and Shelley had investments on either side of the line dividing what we now call "literature" and "science." Though Shelley drew on

Parkinson's account of fossils, which was partly based on Smith, and Smith him-self wrote verse, their works are politically and generically too remote from each other to suggest direct influence.[4] Precisely because direct influence is unlikely, the striking parallels between Smith's and Shelley's fossil archives illuminate a signifi-cant cultural moment: the new, comprehensive geological map—whether visionary (Shelley) or practical (Smith)—reconfigures the established analogy between the "history of the earth" and the "history of nations" to produce a socially constituted materiality.

Applied Geology and the Rock Record

William "Strata" Smith (1769–1839) is the first English interpreter of a rock record still recognized as such by geologists, thanks to his major discovery: the importance of "guide" fossils for correlating rock strata. When a particular fossil organism is found mainly in various outcroppings of the same rock type, it can be assumed that those outcroppings originally comprised one continuous stratum, deposited at a time when that organism thrived. Smith writes:

> By the help of organized fossils alone, a science is established with characters on which all must agree. . . . The organized Fossils (which might be called the antiq-uities of Nature) . . . are so fixed in the earth as not to be mistaken or misplaced; and may be as readily referred to in any part of the course of the stratum which contains them, as in the cabinets of the curious.[5]

Smith joins the logic of orderly collection and the motives of improvement to earlier, more speculative notions of a rock record, such as Hutton's and Davy's. Smith's technique of stratigraphic correlation, traditionally emphasized by histo-rians of geology, enabled him to produce his historic *Geological Map of England and Wales* (1815), recently in the limelight owing to Simon Winchester's widely re-viewed new biography, *The Map That Changed the World* (2001).[6] I am concerned with Smith's largely neglected writings, but Winchester's emphasis on the map is convincing, and Smith made other important innovations in the graphic depiction of the rock record.

One of Smith's numerous tables provides a striking analogue for Shelley's de-scription. The image (Figure 2.2) represents the earth, like Shelley's drama, as both a stratigraphically organized mine of useful material and a subterranean archive of successions of prehistoric creatures. This table (published in 1817) is one of several remarkably comprehensive geological surveys of the island deriving from Smith's 1815 map, the first systematic representation of large-scale stratigraphic succession in our modern sense (cp. Figures 2.3, page 60, and 2.4, page 66). The table is equally encyclopedic in scope, compressing into tabular form the comprehensively transparent rock record Smith had reconstructed from his fossil "characters." The second column names all the English strata in order of superposition, and its heading again emphasizes the logic of collection, which contributes to the narrative structure of the rock record. The first column lists the names of the fossil genera through which the strata become "intelligible"; the extreme right-hand column lists the "products of the strata" through which they become "useful." While Shelley does not trouble himself to specify what substances "make tolerable roads" (see the entry for "cornbrash" in Figure 2.2), he is equally concerned to extrapolate a social and economic narrative through the transparency of the rock record, as I shall argue further below.

Historical geology becomes a study of progress toward the "intelligible and useful," while Nature's improvement of landforms is seen—on both sides of disciplinary boundaries less rigid than our own—as a model for the economic improvement of landscape and a basis for aesthetic categories. While the varying senses of "improvement" complicate this relationship, a fundamental connection between the knowable and the useful seems to underlie any notion—literal or metaphorical—of the rock record.[7] In Smith's version, the rock record emerges as a minutely structured narrative, animated by fossil "characters." The rigorous new distinction between the inorganic and these "organic remains," along with stratigraphic sequence itself, gives the rock record the shape of a narrative; this narrative seems to culminate in human civilization. Charles Darwin was not the first to derive a notion of biological evolution from the history of the earth, but in the early nineteenth century, this evolution remained teleological, as much for Shelley as it did for Smith.

The naturalist's act of reading embodies a hermeneutic sovereignty crowning the evolutionary narrative. Smith's act of reading is at once concrete, relying on his practice of applied geology, and a metaphorically suggestive literary project. He began his study of the strata "while employed in the underground surveys of collieries" and became convinced of their uniform orientation and succession while making his preliminary survey for the Somerset Coal Canal Company in 1796.[8] A manuscript note of that year indicates the relation between this discovery and Smith's pioneering insight into the distribution of fossils: Nature, he observes, "has assigned to each Class [of fossils] its particular Stratum."[9] The historian David Allen stresses the excavations required by large-scale improvements such as mining, canal building, and drainage—in all of which Smith was involved—as the greatest stimulus to the fledgling sciences of stratigraphy and paleontology. Allen also points out that "for the first time," it became possible to "make a living . . . as an out-and-out consultant in [geology], providing the landed gentry with reports on the mineral and soil potential of their estates."[10] The long delay between Smith's initial research and his authoritative publications of 1815–1817 is partly due to the resulting sense that, as Allen puts it, "his data were of great commercial value" (*Naturalist* 50).

Continued on next page

GEOLOGICAL TABLE of BRITISH ORGANIZED FOSSILS.

WHICH IDENTIFY THE COURSES AND CONTINUITY OF THE STRATA IN THEIR ORDER OF SUPERPOSITION;

AS ORIGINALLY DISCOVERED BY W. SMITH, *Civil Engineer;* WITH REFERENCE TO HIS

GEOLOGICAL MAP of ENGLAND AND WALES.

ORGANIZED FOSSILS which Identify the respective STRATA.		NAMES of STRATA on the Sections of the GEOLOGICAL COLLECTION	COLOURS on the MAP of STRATA	NAMES in the MEMOIR and the PECULIARITIES of the Strata.	PRODUCTS of the STRATA.
Nobuca, Rockshirita, Fusus, Crabla, Small Teeth, Crabs Teeth, small Bones	*Plains*	London Clay		London Clay forming Highgate, Harrow, Shooters and other detached Hills	Septarians from which Parker's Roman Cement is made
Marine Turbo, Echinoderm, Cardita, Venus, Ostrea		Clay ... Sand		Clay or Brickearth with interspersions of Sand and Gravel	St. Building Stone in all this extensive District but Abundance of Whetstone which make the best Bricks and Tile in the Island
		Crag ... Sand		Sand & light Loam upon a sandy or absorbent Substratum	Potters Clay, Glass Grinders Sand and Loam and Sands used for Various Purposes
Flint, Thyronia, Ostrea, Echini — Plagiostomum	*Chalk Hills*	Chalk Upper / Lower		Chalk	Flints the best Road Materials
Terebratula, Teeth, Palates — Plagiostomum		Green Sand		Green Sand parted by the Chalk	Good Lime for Water Cement
Funnelform, Reynella, Fensa, Onimu, Perivia, Terebratula, Echini					Firestone and other soft Stone sometimes used for building
Belemnite, Ammonites		Brickearth		Blue Marl	
Turritella, Ammonites, Trigonia, Dvina Wood	*Clay Vale*	Portland Rock ... Sand		Portland Stone, Kentish Rag and Limestone or the Vales of Pickering and Aylesbury	The best Quarry and building Stone abovenamed in the Series. Kimmeridge Coal
Trochus, Nautilus, Ammonites in Masher, Ostrea in a bed Bouer		Oaktree Clay ... Sand		Iron Sand & Carstone which in Sorry and Bedfordshire contains Fullers Earth and in some Places Ochre and Glass Sand	Fullers Earth, Ochre and Glass Sand. Green Lime used on these Sands in Sussex and Yorkshire
Various Madrepores, Melania, Cidaris, Echini and Spines		Coral Rag and Pisolite			
Isobranchia, Ammonites, Ostrea		Clunch Clay and Shale		Dark blue Shale producing a strong Clay Soil chiefly in Pasture in North Wilts and Vale of Bedford	
Ammonites, Ostrea		Kelloways Stone			Makes tolerable Roads
Modiola, Cardia, Ostrea, Avicula, Terebratula		Cornbrash		Cornbrash A thin Rock of Limestone chiefly arable lying in Clay	
Pectens, Teeth and Bones, Wood	*Stonebrush Hills*	Sand & Sandstone		Forest Marble Rock thin Beds used for rough Paving and Slating	Coarse Marble, rough Paving and Slate
Pear Encrinus, Terebratula, Ostrea		Forest Marble			
Madrepore, Cardia		Clay — over the Upper Oolite			
Modiola, Cardia		Upper Oolite		Great Oolite Rock which produces the Bath Freestone	The finest Building Stone in the Island for Gothic and other Architecture which requires nice Workmanship
Madrepores, Birds, Nautilus, Ammonites, Pecten		Fullers Earth & Rock			
Ammonites, Belemnite as in the under Oolite		Under Oolite		Under Oolite at the Vicinity of Bath and the midland Counties	
		Sand			

Minerals Fluids for Coal

Such as which Lime is surely used as a Manure

Figure 2.2.—*continued*

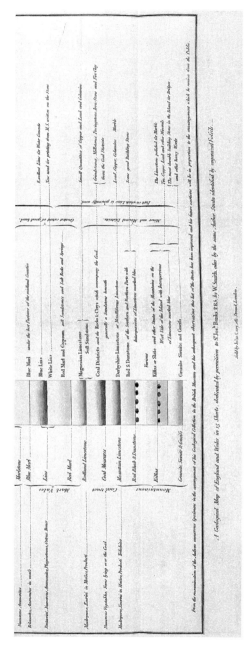

Figure 2.2. William Smith, *Stratigraphical System of Organized Fossils* (London: E. Williams, 1817), Geological Table of British Organized Fossils. Courtesy of the Linda Hall Library, Kansas City, MO.

SYNOPSIS OF GEOLOGICAL PHENOMENA.

SOURCES OF EVIDENCE.	DEDUCTIONS.	RESULTS.	REMARKS.
Fishes in abundance .... Petrified shells are' so filled .... Crystals blunted prior to union in masses .... Shells in rocks not crushed, ... —— by hardening of matter .... —— none in open joints and veins, ...	indicate that .... as to evince minute solution .... show aqueous action.......... therefore rocks hardened quickly .... were fixed, ... therefore joints and veins opened subsequently to their fixation, ...	Water prevailed. Polarity of atoms. 1. Crystallization of particles. 2. { Aggregation, 3. { induration, and cementation of mineralized masses in the *Stratification.*	
Mineral veins obviously were cavities—are filled with spar and ore—spar, the principal, filling. This therefore, traced through cavities of all sorts, in all the rocks, from chalk to mountain limestone, including cavities in fossil shells, and those caused thereby in blocks of stone, into which no gross matter could enter	shows that all these, and thus by analogy, that	*Mineral Veins* were filled by segregation.	Where the mineral veins are cavernous, there, and there only, all the fine crystalline cabinet specimens of minerals occur.
The trade winds and sea currents .... Which force at the completion of the earth .... By the spheroidal figure of the earth	are existing effects of the earth's centrifugal force. arrived at its maximum effect. water as well as land is 13 miles higher under the equator than at the poles. } General Action	This uplaying fluid action was therefore the origin of HILLS which with passes thro' the liquid matter of all strata being successively kept open was the origin of VALLEYS.	
Evidently such force was deflected by primitive rocks, and further as the matter formed, locally—the elevating force remaining greatest at greatest depths of water	therefore the force which caused that general elevation under certain circumstances was adequate to the casual elevation of land, 3 or 4 miles higher than the ordinary level.		
By the remains of land animals mixed up with water-worn stones	We ascertain that the	Earth was dry and inhabited.	
By the bouldered stones every where scattered over the earth's surface, ...	There has been water in action.	} THE DELUGE.	
By the fossil shells in those boulders, identified with those in the stratified rocks, ...	We ascertain the way of action		
By the height to which the boulders and sea-shells have been raised, ...	We get the force of action and height of the water.		

Continued on next page

Figure 2.3.—*continued*

ILLUSTRATIVE EFFECTS OF THE DELUGE.

By alum-shale, organised fossils, those of coal, and mountain limestone, and boulders from all the rocks northward, in abundance,	The effects of a great current from the N. are obvious on the Yorkshire coast.	The first rush of water was by sea from the North.
By the same,	with the like effects,	Down the vale of York, from N.
By the absence of alum-shale fossils in the vale of Pickering,	Filey cliff was not surmounted, which gives the height of	First rush of water about 200 feet.
By bays being filled up, and low places inland, as at	Staiths, Whitby, Scalby, Scabro; and all Holderness,	by westward uplaying.
By whinstope, porphyry, conglomerates, jasper, etc., etc.	on Suffield heights, etc.	From the N.
By Shapfell granite, mountain limestone, etc.	Cleveland hills, Lestingham, Suffield hill, etc., to coast of Holderness,	From the N.W.
By sea shells on the Lancashire coast, By seas shells under 20 feet of gravel, By rounded chalk and flints,	1000 feet high in Snowden mountains, on side of ditto, from north of Ireland Vale of Taunton and below Bristol, on the hills near Bath, Cricklade common,	With wonderful uplaying, from N.W.
By flints from the chalk hills, By bouldered chalk and flints, far in By flints,	Northamptonshire, Rutland and Huntingdon Ashby de la Zouch, and vale of Trent,	Currents from S. and S.E.

Figure 2.3. Synopsis of Geological Phenomena by William Smith, in Thomas Sheppard, *William Smith: His Maps and Memoirs* (Hull: A. Brown, 1920). Courtesy of Widener Library, Harvard University.

Smith's stratigraphic system, based on the "guide" fossils, thus responds to economic impetus, but is also informed by the need for an accessible science and for moral justifications of utility. Its empirical rigor, motivated by economic interest, generates a uniquely complete and transparent archive whose clarity in turn vindicates utility, revealing an inexhaustible wealth that is economic as well as moral, because it records a teleology of increasingly "organized" forms of life.

Smith's accessible new science—a study of "characters on which all must agree"—makes geological order identical to human ordering. He recalls becoming convinced of the earth's legibility during his early assignments as a surveyor: "it was the nice distinction which those similar rocks required, which led me to the discovery of organic remains peculiar to each Stratum" *(SIOF* 2). Smith claims that his system will bridge the gap between theory and practice because it is a science requiring only practical skills, operating on a self-evident and objective order. Fossils are arranged as perspicuously among the strata as in "the cabinets of the curious": to the trained eye, whether the landlord's or the artisan's, the rocks provide a transparent chronicle of the deposition of strata, and the guide fossils Smith identifies permit an exact rendering both of mineral history and of earth history. He returns to the analogy of the cabinet in 1817, promising readers the ability to "search the quarries of different Strata . . . with as much certainty of finding the characteristic Fossils of the respective rocks, as if they were on the shelves of their cabinets" *(SSOF* v). The transparency of the rock record is such that its intrinsic order can be confirmed by matching abstracted specimens with specimens still embedded in their respective strata. Any fossil specimen confirms the uniformity of stratigraphic succession: "the Geologist is thus enabled . . . to fix the locality of those previously found . . . and to find in all former cabinets and catalogues numerous proofs of accuracy in this mode of identifying the Strata." Confirming these suggestions, Simon J. Knell argues that "the new science of geology found its focus . . . [in] fossils and their utility" largely thanks to Smith, who "invented the geological museum."[11]

The logic of orderly collection, once again, proves identical with the internal logic of stratified rock, and geological order seems fully revealed (Figure 2.3). Stratigraphy, Smith writes,

> must lead to accurate ideas of all the surface of the earth, if not to a complete knowledge of its internal structure, and the progress and periods of its formation; for nothing can be more strongly and distinctly marked than the line which separates the animal from the vegetable fossils, and the courses of numerous strata, which are designated by these and other characters, the most intelligible and useful. (*Memoir* 6)

Like Davy's sublime cosmological narrative, the intimation of perfect and comprehensive geological knowledge here legitimates the utility of the rock record. In

Smith's version, such complete knowledge is obviously, rather than "apparently," "within reach of our faculties," and in claiming the entire planet as the territory of his stratigraphy, he simultaneously plants the flag of "improvement" on all the planet's mineral resources. The idea of the globe as a geological entity owes as much to these economic concerns as to Alexander von Humboldt's discovery that the same strata succeed each other on widely scattered land masses.[12] The planet's geological uniformity promises the triumph of European economic geology as globally profitable science—a promise seemingly confirmed by the eventual Englishing of whole periods of the planet's history (e.g., Cambrian, Devonian).

Smith's writings give voice to a fundamental shift in the cultural status of rocks. Hutton's *Theory of the Earth*, too, represented the deformation of strata and evolution of rock types as a legible text registering the motive forces of earth history. But Hutton appealed for evidence of these forces to older aesthetic categories of sublime power and alien physicality, and his reading of the rock record was derided—if partly on political grounds—as obscure and fantastic. Smith's theory reconfigures the strata as a text on which biological order inscribes itself; the rocks become a comprehensive archive of natural history. His recurring appeal to "characters" makes the analogy of text and strata more concrete. Smith's work epitomizes the new tendency to narrativize a scientific distinction between organic and inorganic substance. The intrinsic importance of rocks as a mysterious and alien territory decreases in proportion as they are found to be instrumental for a study of the more accessible organic past. Once the otherness of rock becomes institutionalized as a chemical distinction, it can be integrated into a hierarchically ordered cosmology. The aestheticized otherness of rocky landscapes remains an important topos in the era of Smith, Cuvier, Parkinson, and Buckland, but it appears side by side with an increasingly orderly rock record.[13] Smith's history is more strictly empirical, but also theologically more palatable (for the period) than Hutton's, and full of infectious wonder at the sheer quantity and distinctness of the facts shown by the fossil record. The definition of "intelligible and useful" strata provides a geological context for fossil organisms that ultimately makes intelligible the formation of the earth, as purely inorganic geology had not done before: "each layer . . . must be considered as a separate creation; or how could the earth be formed *stratum super stratum*, and each abundantly stored with a different race of animals and plants[?]" (*SSOF* vii). Smith's religious rhetoric illustrates the symbiotic relationship, too often reduced to simple antagonism, between religious orthodoxy and scientific discovery.

The new geological context supplies natural history with a teleology as certain as stratigraphy: "there seems to have been one grand line of succession, a wonderful series of organization successively proceeding in the same train towards perfection" (x). Because it preserves specimens perfectly and sequentially, Smith's archival rock record permits the reconstruction of earth history as such a teleological narrative. The archive holds "treasures of an ancient deep, which prove the antiquity and watery origin of the earth; for nothing," Smith argues, "can more plainly than the Zoophites evince the once fine fluidity of the stony matter in which they are enveloped" (ix). Zoophites and shellfish must then be the most primitive animals, having existed before there was dry land at all, "as they have entered into the composition of a large proportion of the solid parts of the earth" (x). The rock record here becomes the substance of history, as well as its text, because it embodies the successive creations it records. The teleology of increasingly perfect "organization" subsumes a progress from the liquid to the solid state. Fossil organisms seem the animating principle behind the geological evolution of a planet that eventually comes to sustain large land mammals like human beings. The strata themselves, rigorously distinguished as inorganic, become a neutral repository for vivid biological inscription, paradoxically permitting the virtual animation of specimens contained by them. If "many strata" are in fact composed chiefly of these remains, this underscores their function as a medium. They preserve fossil organisms "without violence" and display them "entombed . . . with all the form, character, and habits of life" (x–xi).

Not only this teleology, but also the local and economic nature of Smith's geology, lead him to a quasi-Neptunist emphasis on marine deposition as the mechanism of rock formation.[14] The theory of "watery origin" best accounts for the stratified sedimentary rocks that make up the vast majority of his geological map of the island, and particularly explains the perfect preservation of England's wealth of fossils and minerals. The most obvious link between these two is coal; Smith explains that "vegetable impressions particularly define in the collier's shaft the approach to coal." His historical interpretation emphasizes the aesthetic and moral profit afforded by the fossil record: "Thus endless gratification may be derived from mountains of ancient animated nature, wherein extinct plants and animals innumerable, with characters and habits distinctly preserved, have transmitted to eternity their own history, and the clearest and best evidence of the earth's formation" (x). The recurrence of economic language in these elucidations of the rock record, so often intertwined with the aesthetic in the rhetoric of improvement, suggests that moral profit is organically linked to financial gain: "organized fossils are to the naturalist as coins to the antiquary; they are the antiquities of the earth; and very distinctly show its gradual regular formation, with the various changes of inhabitants in the watery element" (ix–x). Smith's rhetoric of superabundance, above all, calls attention to the economic subtext of this history of the earth. The passages I have quoted conjure up an earth "abundantly stored" with "treasures of an ancient deep," "mountains" of "innumerable" fossils, an "immensity of animal and vegetable matter," and similar magnitudes. This wealth is metaphorical and literal at once, not only because of the connection between coal and vegetable matter, but because it is discovered under a utilitarian premise. Fossils, Smith insists, "are not the sports of nature . . . , but they must . . . have their use" (vii). With a stock of natural resources seeming to rise effortlessly—at home and in the colonies—to meet the needs of industrial expansion (one might cite the threefold increase in coal production between 1800 and 1830), all of nature falls under the rubric of inexhaustibility.

Smith's 1815 *Memoir* most explicitly reveals the economic context of his stratigraphic system of fossils. This text, which accompanies his geological map of England and Wales—the first of an entire nation—sets out to demonstrate that such a map is an object of "national concern": "The wealth of a country primarily consists in the industry of its inhabitants, and in its vegetable and mineral productions; the application of the latter of which to the purposes of manufacture, within memory, has principally enabled our happy island to attain her present pre-eminence among the nations of the earth" (*Memoir* 1–2). A geological map of the nation can evidently increase the efficiency with which resources are exploited. Figure 2.4, a cross-section derived from the map, indicates the national scale of Smith's project and is more effectively reproduced in this space. His *Memoir* goes on to detail the useful applications of minerals in every conceivable branch of the economy, producing the kind of catalog later schematized in the table of fossils (see Figure 2.2). "Great benefits may accrue both to the landed and commercial interests of the country" from a knowledge of stratigraphy, as he urges in concluding this litany of economic opportunities (4). The regularity of stratigraphic succession, as proven by the distribution of fossils, guarantees the economic efficacy of the map, which must enable improvers accurately to predict geological conditions.[15] This reliable bottom line is certainly one reason why Smith recommends his master text to the "virtuoso," predicting that "his own house will be the best school of Natural History for the younger branches of the family . . . science will become more general. . . . No study, like that of organized Fossils, can be so well calculated for the healthful and rational amusement of youth" (*SSOF* v–vi).

One point of intersection between such applied geology and literary culture lies in the moral dimension of "improvement." Like Smith's, the career of Mary Anning (1799–1847) as a prospector and dealer in fossils at Lyme Regis shows the economic impetus to empirical rigor. Anning's contributions to academic paleontology, and the moral interest of her life as a simple orphan who made such a career for herself, also illustrate the pedagogical function of the rock record, the moral profit that accompanies economic gain and aesthetic experience. The pedagogical value of natural history—claimed by Smith for stratigraphic paleontology in particular—is one cultural legacy conditioning the emergence of the paradigm of a rock record. This value derives in part from the culture of sensibility, with Ann Radcliffe's novels providing one pertinent, if perhaps surprising, example. The naturalizing of Radcliffe's heroines—who scrutinize and narrativize landforms to

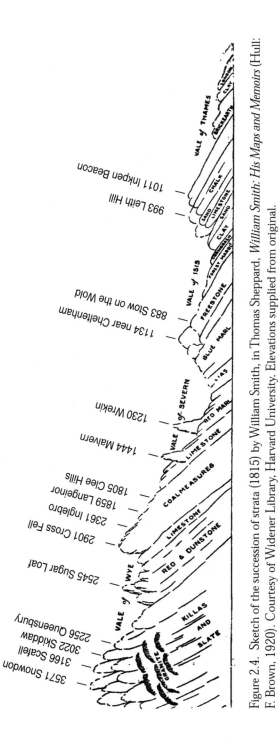

Figure 2.4. Sketch of the succession of strata (1815) by William Smith, in Thomas Sheppard, *William Smith: His Maps and Memoirs* (Hull: F. Brown, 1920). Courtesy of Widener Library, Harvard University. Elevations supplied from original.

derive the sort of moral education that must be acquired by *reading*—helps to account for the currency of the old trope of the "book of nature" around 1800, as well as providing the gendered paradigm that informs Anning's heroism as a "handmaid of geological science."[16] For Smith, too, the rock record is fundamentally a moral text, and individual response to it is an index of moral and religious feeling: "If these animals and vegetables had only to live and die, and mark respectively the sites of their existence in the mass of matter which now forms the earth, they have had their use, and will forever remain indefaceable monuments of that wonderful creative power" (*SSOF* xi).

This evolutionary narrative, with the act of reading that crowns it, is another point of intersection for the geological and the literary. Literary texts attend increasingly to the telos of this evolution, the connection between natural and human history. The rock record in these cases is metaphorical in a different sense from the geologist's chronicle of literal geological events. While the geologist's rock record authorizes human agency, particularly the economic agency of prospecting and other forms of "improvement," the "metaphorical" rock record speaks directly of human agency. In Wordsworth, for instance, it provides a natural model for the poetic practice of inscription, while for Novalis the history of the earth is its spirit, its identity with human agency. The scientific premises for the rock record are intrinsically of literary interest, especially the strong distinction between organic and inorganic matter. The earth's interior in *Prometheus Unbound* becomes an archive much like Smith's, in which the inorganic is illuminated by organic inscription, though in Shelley's version human cultures also participate in this inscription. At the same time, the older notion of "primitive rocks" persists as an impenetrable inorganic remainder, again a figure with literary potential, against which human agency can be usefully set off.

It becomes more difficult to maintain these distinctions between "literary" and "geological" metaphors in light of the interpenetration of early earth science and the sphere of letters. This relationship is more plainly visible in German Romanticism, since there was an established state mining bureaucracy in Germany in which both Goethe and Novalis received training.[17] Shelley's poetry exhibits the more localized influence of geological reading as well as strong thematic affinities to earlier geologists such as John Whitehurst, whose account of apocalypse strikingly resembles the illuminated earth in *Prometheus Unbound*. I have spent considerable time with Smith's exactly contemporary texts, however, because these bring out the narrative sequence and the economic context that are fundamental to Shelley's account of the rock record as the substance of history. The pronounced differences between Shelley and Smith, on the other hand, reflect the gradual formation of disciplinary boundaries; it thus becomes clearer that the "rock record" is *both* a literary and a scientific metaphor and that the aesthetic provides language necessary to imagine human control over resources, without being reducible to this economic function.

The Secrets of the Earth's Deep Heart

Romanticism's illegible rocks have loomed large in the scholarship, at least partly because they pose such a hermeneutic temptation to critics. The blankness of the Simplon Pass in Wordsworth has prompted numerous influential readings, while the "naked countenance of earth" that Shelley sees in Mont Blanc is often clothed with signification in a critical move exemplified by Frances Ferguson's "What the Mountain Said."[18] But there are plenty of legible rocks in Romantic poetry, as in the "fossil scene" in Charlotte Smith's *Beachy Head* and the critique of applied geology in Wordsworth's *The Excursion*, or—in a different sense—Wordsworth's many "inscription" poems.[19] Paolo Rossi has shown that the practice of reading rocks as historical documents dates back to the late seventeenth century. Charlotte Smith and Wordsworth, in fact, are somewhat ambivalent readers of rocks because they find them too thoroughly inscribed with scientific theory and historical narrative. At the end of this essay, I turn to John Whitehurst, whose use of geology in 1778 to vindicate "the great antiquity of arts and civilization" shows how deeply rooted was the analogy between "the history of the earth and the history of nations" (in Rossi's phrase). This analogy, rather than the Wordsworthian attitude of "Mont Blanc," moves Shelley to adopt the image of the archive for his representation of the liberated earth. *Prometheus Unbound* is thus affiliated not only with English applied geology (such as that of Whitehurst and William Smith) but with the continental tradition identified by Rossi and its more recent exemplars in Germany.

German Romanticism provides some of the closest analogues for Shelley's liberated earth. Abraham Gottlob Werner formulated the eighteenth century's most systematic and influential model of the rock record, though his model of deposition was based on the density rather than the age of the strata. Werner's institutional forum, the Freiberg Academy of Mining, helps to account for his tremendous European influence—much greater than Smith's in both geology and poetry. Novalis was probably the most literary among the civil servants Werner trained for the state-run mining industry. Novalis's poetic goal of a liberated nature is deeply akin to Shelley's in *Prometheus Unbound*, but also shows clearly the influence of Werner and economic geology. Novalis proposes, in *Die Lehrlinge zu Sais*, that "natural scientists and poets have always shown themselves, through a common language, as one people." A debate among the "apprentices" and their guests concludes by subsuming scientific knowledge of nature under poetic knowledge, as a humanizing form of domestication.[20] The Temple at Sais has generally been read as an allegory for the mining academy, a reading which makes the teacher or master of the temple a figure for Werner. Another such figure is the old miner in Novalis's novel *Heinrich von Ofterdingen*.

In both works, Novalis strives toward a rejuvenated human relationship to nature, animated by a paradigm of utility drawn from mining: miners understand geology, in his view, because they have an authentic use for the earth's material. Alexander von Humboldt, who studied with Werner shortly before, articulates such a radical form of use-value in his analyses of mineral economy.[21] For Novalis, however, this utility inspires a reverence for nature, which in turn prompts the rocks to yield to miners a knowledge of the historical record that ultimately proves the identity of nature and spirit: "All divine things have a history, and should not Nature, the only whole with which human beings can identify themselves, be included in a history just as human beings are, or have a spirit, which is the same thing?" (*Schriften* 1:99). Novalis's poetic domestication of nature, then, takes mining as its model and proposes to liberate nature by liberating the spirit in the rock record.[22] The rock record in Novalis provides both a tool for locating mineral resources, as in Smith and Werner, and the history that imbues nature with spirit, as demanded by the third traveler in *Lehrlinge*. Mining is poetic geology in the sense that it carries on the *Entwilderung* originating with poetry (1:211). Beyond testifying to the spirit in nature, the rock record for Novalis is the privileged repository of history; in one fragment, he refers to the ordering activity of philosophy as "historical mineralogy" (3:335), and in *Heinrich* (chapter 5), the presence of a studious hermit amidst the fossilized bones of a cave again links natural and human history.

The hermit's cavern with its grottoes full of bones provides one historical link between Novalis and Percy Shelley. In *Prometheus Unbound* IV, Shelley draws on James Parkinson's *Organic Remains of a Former World*, which stages a dramatic succession of fossil fauna in a famous cavern in Gaylenreuth—also the probable basis for the Franconian cavern in Novalis's novel (1:252–66). Shelley focuses less on the competition between poetry and science, perhaps because his scientific background is formed largely by such reading and lacks the practical bent of Novalis's scientific education. Nevertheless, Shelley's "scientific method," like Novalis's, draws on the rock record to construct a vision of liberated nature; like Novalis, he is moved to such reflection by the narrative of zoological succession staged in the bone cave. Shelley's vision calls for a renovation of the globe, restoring the golden age of Edenic harmony that is one of the grand unifying themes traditionally seen as linking Romanticisms. But the resemblance between Shelley's and Novalis's mechanisms for achieving this renovation goes beyond this general thematic affinity. Panthea's vision of the earth in act 4 literally enacts Novalis's protocol for domesticating the earth, which is to make its history legible through spirit.

Act 4 celebrates the psychological liberation of humanity, now "sceptreless, free, [and] uncircumscribed" (3.4.194). As in *Queen Mab* VIII, where the revolution produces a universally habitable earth, Prometheus's liberation here extends to physical nature, and the description of that nature in act 4 becomes a part of the hymn to universal liberation. Panthea and Ione, who function as quasi-choral figures throughout the drama, join the choruses of spirits in act 4 in celebrating renewal as they witness the approach of the Earth and Moon, which will share the majority of the dialogue. I am mainly concerned with Panthea's vision of the earth, especially the second half of her speech which explicates the vision (270–318). The Spirit of the Earth is liberated within the globe, illuminating it from within. The earth appears as a sphere composed of thousands of spheres, all crystalline—solid, yet perfectly transparent, suffused with music and light (238–41).[23] The subsequent explanation of this radiant sphere in terms of recognizable features of the planet, as it was, relies on a notion of the literal and metaphorical opacity of rock (so prominent in Blake) that now dissolves into transparency. The source of light is a star on the Spirit's forehead, sending out shafts of light through and beyond the planet, like spokes on a wheel of which the planet itself would be the hub. These shafts of light,

> as they pierce and pass[,]
> Make bare the secrets of the Earth's deep heart,
> Infinite mine of adamant and gold,
> Valueless stones and unimagined gems,
> And caverns on chrystalline columns poised
> With vegetable silver overspread,
> Wells of unfathomed fire. (278–84)

In this form, the rock record locates mineral wealth more effectively than Smith or Werner could imagine; Shelley's program of liberation, again like Novalis's, calls for natural resources to make themselves perfectly available. Novalis incorporates the exploitation of resources into a poetic program of domestication, but in Shelley's vision the struggle for subsistence (as idealized by the teacher in *Die Lehrlinge zu Sais*) is eliminated entirely. The domestication of nature here entails an absolute human sovereignty, to which resources willingly yield. The Earth, in its longest speech, describes human will as "compelling the elements . . . with a tyrant's gaze" (396–97) and "forcing life's wildest shores to own its sovereign sway" (411, cp. 418–20). The imagery suggests that the Earth is recalling the toppled tyranny of Jove, which Shelley indeed uses similar language to describe. This association undercuts the revolutionary difference between the ancien régime and Promethean "Love," which seems a less "environmentally sensitive" version of the miner's sexualized love for the earth in *Heinrich*. (Compare Alan Bewell's remarks on this topic, pp. 124–25 below.) For Shelley, human liberation entails a complete liberation from economic necessity. The rock record answers this need perfectly by revealing a mineral wealth that is "infinite," "unfathomed," "valueless," and even "unimagined." The geology of the passage is thus an economic geology, a representation of the earth as a fund of resources. The poetry absorbs the environmental attitude of its time and makes it compatible with a poetic reverence for nature (Shelley credits the "divinest climate" and scenery of Italy with inspiring him to write the drama), producing an aesthetic vocabulary and a model of human agency equally amenable to emergent modern science. Shelley represents economic value by means of an economic sublime, which embellishes mineral wealth with an aes-

thetic value (as in the "chrystalline columns") anticipating the strata of human artifacts next illuminated by the Spirit:

> the beams flash on
> And make appear the melancholy ruins
> Of cancelled cycles; anchors, beaks of ships,
> Planks turned to marble, quivers, helms, and spears
> And gorgon-headed targes, and the wheels
> Of scythed chariots, and the emblazonry
> Of trophies, standards, and armorial beasts
> Round which Death laughed, sepulchred emblems
> Of dead Destruction, ruin within ruin!
> The wrecks beside of many a city vast,
> Whose population which the Earth grew over
> Was mortal but not human; see, they lie
> Their monstrous works and uncouth skeletons,
> Their statues, homes, and fanes; prodigious shapes
> Huddled in grey annihilation, split,
> Jammed in the hard black deep. (287–302)

The rock record in *Prometheus Unbound* is monumental. It incorporates
human history more fully than any previous version of the rock record: the transparent earth renders history as an open book. Shelley's imagery of the "ruins of
time" evokes over a century of cosmological speculation; "ruin within ruin," for example, recalls John Dennis's response to the Alps—"Ruins upon Ruins"—under
the influence of Thomas Burnet and physico-theology.[24] The apocalyptic images
of lapsed political power and lost civilizations recall the quasi-archaeological ex-
cursions of earlier geologists such as Whitehurst, while also amplifying the theme
so definitively expressed in "Ozymandias." Once the perception of fossils as documents has blurred the boundary between nature and history, the "ruins of time"
are of a piece with the "ruins of nature," both contributing to "the image of a world
which is . . . the fruit of a slow but inexorable process of degradation and decay,"
as Paolo Rossi puts it.[25] But the possibility of progress also arises from the same seventeenth-century tradition, since fossils can be taken to document a *human* history
of unsuspected antiquity, implying a vaster, open-ended timescale for "nations" as
well as the earth (*Abyss* 16–17). William Smith's writings show how vigorously this
positivism operates in the nineteenth-century context. Analogously, the ruins of
Prometheus Unbound are a source of the moral profit which complements the mineral wealth accompanying them; but rather than celebrating power, they entomb
the tyranny and strife dispatched with Jupiter's fall. The fossilized instruments of
war (290–95), for instance, appeal to geological stability to represent an end to
violence and all sublunary mutability.

The passage uses the techniques of natural history to represent the earth as the substance of a legible history. As in Smith's transparent rock record, the organisms (or artifacts) entombed in the strata compose a large part of their substance, and the mineral component becomes a medium or text on which the order of sentience is inscribed. Shelley's monumental episode particularly suggests the influence of James Parkinson. Parkinson believed that the earth was in large part composed of metamorphosed vegetable and animal substances. He devotes much space to the formation of coal and explains the nature of other minerals through a "Theory of the Petrifaction of Wood" (note Shelley's phrase, "planks turned to marble").[26] Limestone, marble, "calcareous spar," and so forth are simply "the state in which [mollusk] remains were intended to exist in the present state of the globe" (*OR* 2:3–4). The proportion of fossil content, moreover, determines a hierarchy of aesthetic value: "In visiting the mansions of the rich, we shall in general find, that, in proportion to the wealth and consequence of the possessor, will the more solid parts of the building be composed of these remains of animated beings, which lived in a former world" (1:9). Shelley's account of animal remains, in the next portion of the passage, is the most visibly indebted to Parkinson and geologists such as Cuvier, Smith, and Buckland, who revived catastrophism and generated the fashion (roughly 1815–1830) for Deluges and dinosaurs. As John Wyatt puts it, "an aspect of Catastrophism which had special appeal to artists and early paleontologists was the notion of a prehuman state where beast fought beast in a struggle, ending for the losing species in extinction."[27] Such a struggle informs Shelley's account of the ruined cities of prehistoric and even prehuman civilizations. The efficient cause of their destruction appears in the fossil skeletons of giant beasts that lie "over these":

> and over these
> The anatomies of unknown winged things,
> And fishes which were isles of living scale,
> And serpents, bony chains, twisted around
> The iron crags, or within heaps of dust
> To which the tortuous strength of their last pangs
> Had crushed the iron crags;—and over these
> The jagged alligator and the might
> Of earth-convulsing behemoth, which once
> Were monarch beasts, and on the slimy shores
> And weed-overgrown continents of Earth
> Increased and multiplied like summer worms
> On an abandoned corpse, till the blue globe
> Wrapt Deluge round it like a cloak, and they
> Yelled, gaspt and were abolished; or some God

Whose throne was in a Comet, past, and cried—
"Be not!"—and like my words they were no more. (302–18)

The account of fossilized animal skeletons exhibits another kind of monumentality, equally relevant to science and poetry. This passage commemorates the dead, as much as prehistoric events, and paleontology similarly produces epitaphs because it involves long-extinct species. Parkinson himself suggests that "many enormous chains of mountains are vast monuments in which these remains of former ages are entombed" (*OR* 1:8; cp. Erasmus Darwin, *The Temple of Nature* 4.450). (Parkinson's "wood changed into marble" no doubt appeals to Shelley partly because marble is the right material for a monument.) Shelley's account can offer a more spectacular concentration of remains because it is not localized, but Parkinson's account of such "monarch beasts" is also inspired by their unparalleled size and destructive power (Figure 2.5). The mastodon, he exclaims, is "one of the most stupendous animals known . . . ; whether we contemplate its original mode of existence, or the period at which it lived, our minds cannot but be filled with astonishment" (3:352). The "inexhaustible accumulation" of such remains provides ample material for the Shelleyan allegory of tyranny ("monarch beasts") deposed, which takes its cue ("and over these") from the layering of these remains. The poem's sequence of events thus gains authority and necessity from the rock record. Parkinson points out, for example, that fossil crocodiles are uniformly found in strata lower than those containing warm-blooded quadrupeds (3:276–77), suggesting a large-scale succession that "explains" local folklore concerning races of "fairies" and "giants" whose conflicts are recorded in the strata (1:4). Parkinson cites a passage from Cuvier (3:401) that predicates the total annihilation Shelley requires for this passage, but disregards the implication of a sequence of catastrophic floods because of his investment in the literal truth of Genesis (449–50). Since the religious rhetoric, as in Smith, accompanies economic incentives (e.g., xii), it seems likely that Shelley's use of Parkinson is somewhat ironic.

The question of direct influence is in any event less interesting than the compatibility of this semi-popular scientific discourse with Shelley's project. As Parkinson points out, a study of the stratigraphic relations among the substances in which fossils occur is necessary to reconstruct a historical record. Shelley loosely adapts this idea of a rock record in order to represent natural history as a discernible sequence of tyrannies, infestations of the earth finally arrested by a diluvial annihilation of such changes, strikingly represented as a process of organic decay. "Monarch beasts" are thus mere parasites upon the corpse of the earth until it is cataclysmically renovated by Prometheus's liberation (*PU* 4.313). Parkinson observes that the prehistoric times to which fossils can be dated are "so remote" and the evidence "so slight" that "the majority" of the theories he summarizes in the course of the work "rather resemble the fictions of poets, than

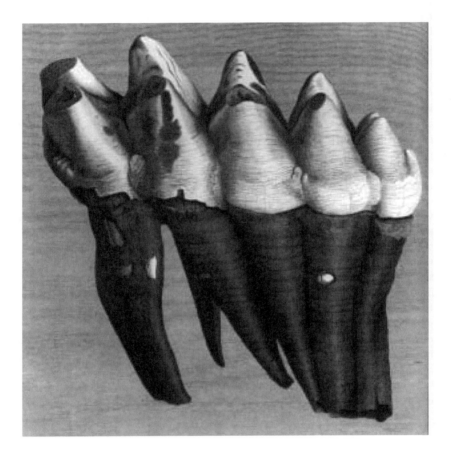

Figure 2.5. James Parkinson, *Organic Remains of a Former World*, 3 vols. (London: J. Robson, 1804–11), Vol. III, Frontispiece: The Back Grinding Tooth of the Mammoth or Mastodon of Ohio. Engraving by S. Springsguth. Courtesy of the Linda Hall Library, Kansas City, MO.

the reasonings of philosophers" (*OR* 1:13–14). Shelley's theory is no exception, and its breathtaking imagery and scope help to explain why Parkinson subtly offers the "poetic" quality as a selling point. But Shelley uses actualism just as impressively as catastrophism in the image of Prometheus chained to his rock in act 1. A part of the curse that forms the centerpiece of this act involves a provocation of Jupiter: "Let alternate frost and fire/Eat into me, and be thine ire/Lightning and cutting hail and legioned forms/Of furies, driving by upon the wounding storms" (268–71). These lines identify environmental upheaval as Prometheus's punishment—hence the necessity of abolishing it in the liberated world. Prometheus, meanwhile, becomes a living part of the rock record, receiving the impress of the forces gradually wasting the side of the mountain to which he is chained.

Panthea's concluding exclamation—"Be not!"—inverts the divine fiat, concentrating the destructive force accumulating throughout the passage into a single point. Here the apparatus of natural history serves the undoing of natural history, the abolition of change and decay, and the liquidation of history itself. The reference to the comet (4.317) appeals to much earlier theories offering the approach of a comet as the efficient cause of the Deluge (this explanation was also revived by Humphry Davy). Shelley acknowledges the disjunction between modern diluvialism and physico-theology by means of the "or" in line 316; he prefers the older theory not for the strength of its professed explanation but because the passage requires a more literal divine intervention. This explanatory equivocation is one of the first signs of a literature self-consciously diverging from science; there are other examples in Romantic poetry of such pairings of alternatives that require, from a scientific point of view, mutually exclusive explanations.[28] The god's command "abolishes" the behemoths and, as a speech–act, terminates the speech of Panthea, the narrator here. Evolutionary succession—from demiurges to one prehistoric animal population to another—also terminates with this verbal Deluge; since the Earth's present-tense speech follows Panthea's narration, we know that human beings, with their contemporary fauna, represent the end of natural history. As Parkinson observes, "the last and highest work appear[s] to be man, whose remains have not been numbered among the subjects of the mineral kingdom" (3:455). Using geological metaphors to reveal it as an arrestable sequence of tyrannies, Shelley's vision liquidates human history as well. The illumination of the rock record completes that history by making available the resources at issue, along with psychological factors, in the struggles comprising history.

Shelley's passage incorporates several important functions of the rock record established by precursors as various as William Smith and Novalis. On the one hand, Shelley's illuminated rock record achieves the transparency anticipated in Smith's claim that "encouragement only is wanting in this important branch of Natural History to unravel the mystery and simplify our knowledge of all the terrene part of the creation" (*SSOF* vii). A wealth of positive knowledge is the least part of the profit yielded by this rock record. On the other hand, a renovative "Love," as the Earth puts it, "interpenetrates my granite mass" (370). As in *Die Lehrlinge zu Sais*, a universal love lifts the veil of exteriority and elevates inanimate nature to the status of a humanized other: "Heaven, hast thou secrets? Man unveils me, I have none" (423). The earth appears, as in Novalis's work, as a rock record on the largest scale, domesticated through the legibility of this record, which provides evidence of its spiritual nature; or, as Shelley has it, domesticated along with nature, through the action of spirit. D. J. Hughes argues that "it is the very structure of things Shelley would dissolve through the action of mind upon matter."[29] The central image of such a recalcitrant materiality is of course the naked rock to which Prometheus is originally chained, "black, wintry, dead, unmeasured; without herb,/Insect, or beast, or shape or sound of life" (1.21–22). In act 4, Shelley's prophetic characters "measure" and overcome that resistance through the vision of a perfectly habitable world. As in the postrevolutionary world of *Queen Mab*, "all things are void of terror" (8.225), purged of deserts, polar wastes, boundless oceans, tempests—all hostile environments, but also, curiously, the poetic resources of the sublime.

The naturalist John Whitehurst (1713–1788) is one earlier practitioner of the spectacular style or geological sublime who seems to share very closely Shelley's vision of a monumental, dystopic earth history. His is an apocalyptic vision, like Shelley's, that constructs apocalypse as a liquidation of natural resources. Whitehurst's catastrophe literally creates the "infinite mine" illuminated by Shelley's apocalyptic light: his "subterraneous convulsions" have the economic function of creating the stratigraphic discontinuities in which useful minerals are deposited, while inundations of "subterraneous fire" not only produce coal deposits but also envelop an astonishing variety of remains, many of them human artifacts. In combining this quasi-archaeological project with the economic one into a "subterraneous geography," Whitehurst establishes the rock record as the fundamental text of history. History thus guarantees the superabundance of resources; the substance of nature equals the substance of history, its domestication disrupted only by a catastrophe that ultimately restores the earth and reveals its plenitude. The parallel of a renovative Deluge—an ironic piece of theological orthodoxy on Shelley's part—accompanies the parallel narratives of successive depopulation in Whitehurst and Shelley, and both parallels inhabit the semantic field of "revolution," the politics of which is more central to *Prometheus Unbound*.

Whitehurst takes the appearance of geological disorder as the point of departure for his theory of "subterraneous convulsions," inspired by the rugged countryside of Derbyshire, where he lived (see Figure 2.6). Remarking on the ubiquity of "craggy rocks and mountains, . . . steep, angular, and impending shores, [and] subterraneous caverns," he explains that "these romantic appearances" are the effect of "some tremendous convulsions, which have thus burst [the] *strata*, and thrown their fragments into all this confusion and disorder."[30] At this point Whitehurst's theory resembles the physico-theology (especially Thomas Burnet's) of almost one hundred years before. But its stronger empirical basis appears in the "illustration" of the theory, which examines evidence of environmental upheaval, including the recent catastrophes of Lisbon and Vesuvius, in order to establish, by analogy, the magnitude of the "great revolutions" occurring "anterior to history" (66). Whitehurst's scientific apparatus includes these spectacular anecdotes and his Derbyshire field observations, as well as a sound knowledge of contemporary physics and chemistry. He is probably the first English writer to appeal to the rock record for scientific testimony in the modern sense, chiefly in order to prove "the existence, force, and immensity of subterraneous fire" (115). In these respects, he anticipates both James Hutton's Plutonism and Smith's applied stratigraphy.

Like Shelley as well as these later geologists, Whitehurst allegorizes the stratigraphic paradigm of a rock record to derive historical evidence of "a period indeed much beyond the reach of any historical monument, or even of tradition itself" (257). The rock record expands into a dimension that allows it to recover the human history of an antediluvian world, despite its disappearance from "human record": "the history of that fatal catastrophe is faithfully recorded in the book of nature, and in language and characters equally intelligible to all nations, [and] therefore will not admit of a misinterpretation." Such notions of clarity and transparency are later taken up by both Smith and Shelley, in their differing idioms, but Whitehurst's earlier account makes clear that these idioms share common cultural sources. The *Inquiry* makes an apt bridge between the early-nineteenth-century discourse and the philosophical tradition identified by Rossi, for whom recognition of the "dark abyss of time" coincided with reflection on the "boundless antiquity of 'nations'" (*Abyss* x). Rossi cites a number of late-seventeenth-century thinkers who argue that just as fossil species have disappeared, "entire cultures" may have been swept away (15), and fossils may be found to "document" deliberately obscure ancient myths of human origins (17). These views are clearly present in Whitehurst, indicating how much the inherited narrative forms of geology contributed to its new industrial applications.[31] A key piece of analogical evidence for Whitehurst's "revolutions" is that whole towns are erased by volcanic eruptions, as in the case of Calloa, Peru, in 1746. His direct evidence of even larger prehistoric cataclysms is necessarily more apocryphal. One detail agreeing with Shelley's account is the presence of ships buried at improbable depths, one of several hints that entire civilizations have been swept away because it implies that "ships were in use at the time of that dreadful catastrophe" (134). Drawing on a wide array of evidence for lost civilizations on several continents—which culminates in a spectacular reading of the Giant's Causeway, a sublime site long associated with Atlantis (258ff.)—

Whitehurst concludes: "therefore to such fatal events, and to the conquests of civ-ilized nations by savage barbarians, we may venture to ascribe the subversion of arts and sciences at sundry periods of time" (272).

The illustration of Whitehurst's theory is also an application of it, a subter-raneous geography "leading mankind to the discovery of many things of great utility which lie concealed in the lower regions of the earth" (178). Whitehurst's overarching Vulcanist–Plutonist thesis leads him not only to explain the Giant's Causeway as the core of an extinct volcano but also to make useful observations concerning the Derbyshire mines. He shows that mineral veins continue under-neath interposing strata of "toadstone" (basalt and dolerite), while the mines below such strata are predictably dry (204–05). The theory makes this insight pos-sible by imagining toadstone (in part correctly) as "actual lava, [that] flowed from a volcano whose tunnel, or shaft, did not approach the open air, but disgorged its fiery contents between the strata in all directions" (197). Whitehurst cites practi-cal application as the primary motivation for his detailed and accurate sections (Figure 2.6), innovative graphic depictions of the strata that are "the ultimate end of subterraneous geography" (210). Accurate stratigraphic representation, he ob-serves, will enable miners and mine owners "to prosecute their mineral researches with more propriety and advantage to themselves and the public" (180). White-hurst's evidence is itself "principally obtained . . . from experienced miners," though verified and augmented by original discoveries he made during his own "subterraneous visits" (181). He touts the sections, as Smith does his, for their ca-pacity to help predict the location of coal and other commodities (204). The goal of an accurate representation of the strata, informed by economic interest, leads here, as in Smith, to the articulation of a rock record of great richness and com-plexity, both mineralogical and literary. The apocalypse or catastrophe on which the larger theory turns has the economic function of creating the spaces for min-eral deposits. And it has the aesthetic function of astonishing readers and connoisseurs of natural history, not just by means of sublime depictions but by establishing the historical legibility of sublime phenomena.

Geology's increasingly practical bent, its concern with physical geography and the "improvement" of land, resonates in surprising ways in Romantic images tradi-tionally viewed as aestheticized representations of nature. Wordsworth, for example, uses the geological categories of "primitive" and "secondary rocks" in his *Guide to the Lakes* to depict a record of improvement inscribed across the landscape by the totality of geological process: "Sublimity is the result of Nature's first great dealings with the superficies of the earth; but the general tendency of her subsequent oper-ations is toward the production of beauty."[32] If Nature itself is an "improver," then the rock record is the monument of a self-domesticating landscape. Whitehurst, Smith, and Shelley all develop versions of this narrative, which is given visual form by the famous image with which I began, from Hutton's *Theory of the Earth*. Though

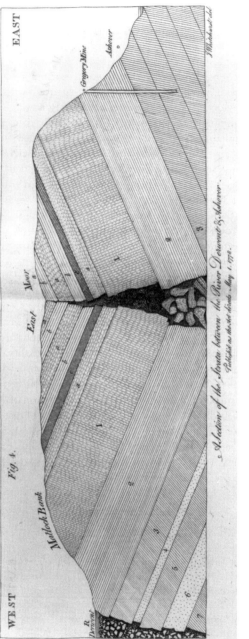

Figure 2.6. John Whitehurst, *Inquiry into the Original State and Formation of the Earth* (London: Printed for the author by J. Cooper, 1778). Figure 4: A Section of the Strata between the River Derwent & Ashover. Engraving from a drawing by White-hurst. Courtesy of the Linda Hall Library, Kansas City, MO.

no English poets had the professional geological training of Novalis or Goethe, broader cultural factors propelled them equally toward the rock record as a subject of poetic elaboration; as a scientific paradigm, the rock record achieved its currency through a geological mastery of literary technique, perhaps most obvious in Whitehurst. The discourse on the rock record works as a whole to condition early industrial perceptions of human agency vis-à-vis the natural world. Shelley's "infinite mine" provides evidence of the wide circulation of Romantic images of mines and mineral resources and, more crucially, of the aesthetic concept of economic agency that accompanies them. The sovereignty over nature that technology seems to afford corresponds to a human sovereignty established by scientific discovery, a sovereignty both hermeneutic and evolutionary. Geology and poetry alike, both still offering in their diverging ways to explain the materiality of nature, find the rock record a valuable instrument of the order required to balance the fascination of sublime disorder, of vast and alien geological forms and processes. It is startling to turn from Romantic rocks that *resist* reading, such as Shelley's "Mont Blanc," or the "huge stone" in Wordsworth's "Resolution and Independence," to the orderly archives the poetry shares with economic geology.

In *Prometheus Unbound*, this order takes the form of a liberated textuality, a revealed identity of textuality and materiality. Tilottama Rajan has argued that "the materiality of narrative and drama, of writing vision into the language of events" in *Prometheus Unbound*, "inevitably defers the re-visioning of history as the phenomenology of mind."[33] If we read Shelley's drama in the context of geology, it becomes apparent that the materiality of the text *is* the materiality of history. Shelley's instrument for the "re-visioning" or transfiguration of history is a phenomenology of the earth, identifying the substance with the text of history. This is not to say that the materiality of *Shelley's* text—its mediation of vision, or verbal and printed form—acts directly on the material conditions of history. Its "beautiful idealisms," rather, are anchored in a historically specific knowledge of the physical world as "intelligible and useful." "Love," says Shelley's Earth, "interpenetrates my granite mass" (4.370), and the resulting narrative absorbs aesthetic and economic forms into one geological, material origin.

Notes

1. I allude to the deconstructive and new historical schools, respectively, of Romantic criticism in the last quarter-century. Rom Harré usefully distinguishes between three kinds of relationships in the study of literature and science: shared content, shared style, and shared "meta-theory." While all three are at issue in this chapter, the paradigm of a rock record fits well into the third category. "What Is the *Zeitgeist?*", *Common Denominators in Art and Science,* ed. Martin Pollock (Aberdeen: Aberdeen UP, 1983), 3.

2. *Shelley's Poetry and Prose*, ed. Neil Fraistat and Donald Reiman, 2nd ed. (New York: Norton, 2002), 277. All subsequent references (by act and line number) are to this edition. The reading "mines" (1.280) is adopted by some other editors. Fraistat and Reiman (278n.) cite Parkinson's *Organic Remains of a Former World* (1804–1811), which Shelley owned; Davy's influence is studied at length by Carl Grabo, *A Newton Among Poets: Shelley's Use of Science in Prometheus Unbound* (Chapel Hill: U of North Carolina P, 1930).

3. *Humphry Davy on Geology: The 1805 Lectures for the General Audience*, ed. R. Siegfried and R. Dott (Madison: U of Wisconsin P, 1980), 77–78.

4. Dennis Dean discusses Smith's importance in Parkinson's *Organic Remains*, vol. 3. See his *Gideon Mantell and the Discovery of Dinosaurs* (Cambridge: Cambridge UP, 1999), 23 and 23n. Smith also cites Parkinson (*SSOF* [defined in note 5 below] v) and Simon Winchester speculates on Smith's possible connections to Davy and John Whitehurst, whose geology is discussed at the end of this chapter. See Winchester's *The Map That Changed the World: William Smith and the Birth of Modern Geology* (New York: HarperCollins, 2001), 94, 123, 224. Simon J. Knell mentions Smith's own poetry in *The Culture of English Geology, 1815–1851* (Aldershot: Ashgate, 2000), 37.

5. *Strata Identified by Organized Fossils* [*SIOF*], *containing Prints on Colored Paper of the Most Characteristic Specimen in each Stratum* (London: W. Arding, 1816), 1–2. The following works of Smith are also cited parenthetically in this chapter: *Stratigraphical System of Organized Fossils* [*SSOF*], *with Reference to the Specimens of the Original Collection in the British Museum: Explaining Their State of Preservation and Their Use in Identifying the British Strata* (London: E. Williams, 1817); and *A Memoir to the Map and Delineation of the Strata of England and Wales, with Part of Scotland* [*Memoir*] (London: John Cary, 1815).

6. See also note 4. Stephen Jay Gould emphasizes preexisting and contemporaneous geological mapping techniques in his review "The Man Who Set the Clock Back," *The New York Review of Books* 48.15 (October 4, 2001): 51–56.

7. In Smith's rhetoric, the three strands of "improvement" identified by Raymond Williams—economic, aesthetic, and moral—are integrally linked and often difficult to unweave. See *The Country and the City* (New York: Oxford UP, 1973), 116.

8. John Farey, "Mr. William Smith's Discoveries in Geology," *Annals of Philosophy* 11 (1818): 360–61.

9. Quoted in Joan M. Eyles, "William Smith: Some Aspects of His Life and Work," *Toward a History of Geology*, ed. Cecil J. Schneer (Cambridge, MA: MIT Press, 1969), 146, cp. 150.

10. *The Naturalist in Britain: A Social History* (1976; Princeton: Princeton UP, 1994), 51. Roy Porter, in "The Industrial Revolution and the Rise of Geology," *Changing Perspectives in the History of Science*, ed. M. Teich and R. Young (Dordrecht: Reidel, 1973), 320–43, has shown how the success of "practical men" such as Smith finally persuaded the scientific elite of the Geological Society to recognize the importance of the fieldwork being done in provincial centers of economic geology. Martin Rudwick, analyzing Smith's much delayed recognition by the Society in 1831, argues that this change of attitude served a certain faction of gentlemen who now wanted to privilege fieldwork. See *The Great Devonian Controversy: The Shaping of Scientific Knowledge Among Gentlemanly Specialists* (Chicago: U

of Chicago P, 1985), 63–68. Knell extensively analyzes the class issues surrounding geological collecting, often touching on the aforementioned aspects of Smith's career. See *The Culture of English Geology*, especially chapters 1, 7, and 9.

11. *The Culture of English Geology,* 305, 74.

12. Smith suspected as much, but evidently had not read Humboldt's geological sketch "prov[ing] the identity of the formations of the two hemispheres," published in 1801, or its successors. See Humboldt's *Personal Narrative of Travels to the Equinoctial Regions of America during the Years 1799–1804*, trans. Thomasina Ross, 3 vols. (London: Routledge, 1851), 1:285. Mary Louise Pratt situates Humboldt's theories in the context of Romantic science and imperialism in *Imperial Eyes: Travel Writing and Transculturation* (New York: Routledge, 1992), 111–43.

13. The pioneering paleontologist Georges Cuvier (1769–1832), recently retranslated by Martin Rudwick, is familiar to many contemporary readers through Michel Foucault, *The Order of Things: An Archeology of the Human Sciences* (1966; New York: Random House, 1971). Lord Byron and William Buckland, who became the first professor of geology at Oxford, were two of the more prominent Englishmen who acknowledged Cuvier's influence. All three participated, in various ways, in an increasingly popular interest in paleontology and in catastrophism that is reflected in Shelley's passage on "the secrets of the Earth's deep heart" and its immediate source, Parkinson. On the chemical classification of rocks, see Rachel Laudan, *From Mineralogy to Geology* (Chicago: U of Chicago P, 1987), chs. 3 and 4; see also ch. 7 on paleontology and stratigraphy.

14. I refer to the thesis of a "watery origin" for all the earth's material in its broad, culturally diffused form. On this point, see Knell's *Culture of English Geology,* 13–14, 18–19. Knell and others point out that Smith was charged with ignorance of the theory of Neptunism as articulated by Abraham Gottlob Werner (cp. *SSOF* vi).

15. Smith's vision of a fusion of science and industry is not, of course, unique for the period. The preface to the inaugural volume of the *Transactions of the Royal Geological Society of Cornwall* (London: William Philips, 1818) evinces some striking parallels to Smith's vision. The epigraph for the volume states the economic implications of geology succinctly: "A knowledge of our Subterranean wealth would be the means of furnishing greater opulence to the country, than the acquisition of the mines of Mexico and Peru" (1).

16. W. D. Lang, "Mary Anning, of Lyme, Collector and Vendor of Fossils, 1799–1847," *Natural History Magazine* 5 (1935): 66–81, here 81. See also Hugh Torrens, "Mary Anning (1799–1847) of Lyme; 'the greatest fossilist the world ever knew,'" *British Journal for the History of Science* 28 (1995): 257–84. For Radcliffe's use of a rock record, see *The Mysteries of Udolpho*, ed. Bonamy Dobrée (1966; New York: Oxford UP, 1980), 602.

17. See, for example, Theodore Ziolkowski, *German Romanticism and Its Institutions* (Princeton: Princeton UP, 1991), ch. 2, or Helmut Gold, *Erkenntnisse unter Tage: Bergbaumotive in der Literatur der Romantik* (Opladen: Westdeutscher Verlag, 1990), for a more detailed study. I draw on both works for the consensus in Novalis criticism cited in what follows.

18. "Shelley's *Mont Blanc:* What the Mountain Said," *Romanticism and Language,* ed. Arden Reed (Ithaca: Cornell UP, 1984), 202–14. Patrick Vincent's "What the Mountain Should Have Said" (North American Society for the Study of Romanticism, September 14, 2000) typifies a new approach that makes the mountain legible by means available to Shelley's contemporaries, such as early theories of glaciation. See Alan Liu, *Wordsworth: The*

Sense of History (Stanford: Stanford UP, 1989), ch. 1, for one influential reading of the Wordsworth passage, incorporating much previous criticism.

19. *Beachy Head* (1807) 368–419, in *The Poems of Charlotte Smith*, ed. Stuart Curran (Oxford: Oxford UP, 1993); and *The Excursion* (1814) 3.50–158, in *The Poetical Works of William Wordsworth*, ed. E. de Selincourt, 5 vols. (Oxford: Clarendon, 1949), vol. 5.

20. Novalis, *Schriften*, ed. Paul Kluckhohn and Richard Samuel, 3rd ed. (Stuttgart: Kohlhammer, 1977), 1:82. My quotations from both *Die Lehrlinge zu Sais* and *Heinrich von Ofterdingen* refer to this edition. The original of the first quoted passage is "Naturforscher und Dichter haben durch eine Sprache sich immer wie Ein Volk gezeigt" (84). For Werner's theory, see his *Kurze Klassifikation der verschiedenen Gebirgsarten* (1786; trans. A. M. Ospovat, New York: Hafner, 1971).

21. For Humboldt's analysis of mineral economy, see, for example, *Political Essay on the Kingdom of New Spain*, trans. John Black, 4 vols. (London: Longman, 1811), 3:104–12. Novalis practiced briefly as a State Inspector of Mines, the profession for which he was trained at Freiberg. In *Die Lehrlinge zu Sais*, the Werner figure remarks that "where people are engaged in complex interaction and conflict with Nature, as in agriculture, navigation, animal husbandry, and in the mines, . . . the development of this [understanding] seems to occur most easily and frequently" (1:108).

22. For the old miner in *Heinrich von Ofterdingen*, to extract metals *is* to liberate them (1:242; cp. the first interpolated lyric, "Bergmannslied"). The rape of nature, rather, as Novalis suggests in *Lehrlinge*, is enacted by theories that subjugate nature purely to human purposes, displacing reverence with instrumental reason. See further Dennis Mahoney, "Human History as Natural History in *Die Lehrlinge zu Sais* and *Heinrich von Ofterdingen*," *Subversive Sublimities: Undercurrents of the German Enlightenment*, ed. E. Timm (Columbia, SC: Camden House, 1992). The oiginal of my previous Novalis quotation is "Alles Göttliche hat eine Geschichte, und die Natur, dieses einzige Ganze, womit sich der Mensch vergleichen kann, sollte nicht so gut wie der Mensch in einer Geschichte begriffen sein oder, welches eins ist, einen Geist haben?" (*Schriften* 1:99).

23. This is the image that Carl Grabo argues is based on Humphry Davy's model of atomic structure, which would strengthen the identification of the earth's material and materiality itself. See *A Newton Among Poets*, 142.

24. *The Critical Works of John Dennis*, ed. E. N. Hooker, 2 vols. (Baltimore: Johns Hopkins UP, 1943), 2:381.

25. *The Dark Abyss of Time: The History of the Earth and the History of Nations from Hooke to Vico*, trans. Lydia Cochrane (1979; Chicago: U of Chicago P, 1984), 37, cp. 4.

26. *Organic Remains of a Former World. An Examination of the Mineral Remains of Vegetables and Animals of the Antediluvian World* [*OR*], 3 vols. (London: J. Robson, 1804–11), vol. 1, ch. 30. Parkinson argues that organic remains undergo "a process of bituminization, by which their conservation is secured, previously to their impregnation with earthy or metallic salts" (3:440). This theory reduces the entirety of geologic process to an archival process for preserving the "organic remains" of vegetables (vol. 1), zoophytes (vol. 2), and animals (vol. 3).

27. *Wordsworth and the Geologists* (Cambridge: Cambridge UP, 1995), 216.

28. Cp. Wordsworth's "Ode: The Pass of Kirkstone," 11–12: "Left as if by earthquake strewn,/Or from the Flood escaped." Or "Mont Blanc": "Is this the scene/Where the old

Earthquake-daemon taught her young/Ruin? . . . or did a sea/Of fire envelope once this silent snow?" (71–74). The comet is a favored catastrophic device in Shelley; cp. "Epipsy-chidion," 368ff.

29. *Shelly's Poetry and Prose*, 1st ed., 611.

30. *Inquiry into the Original State and Formation of the Earth*, 2nd ed. (1786; New York: Arno, 1978), 61.

31. Rossi reads the Romantic period debate about geological time as a recapitulation of the "great dichotomy of deism and materialism," which he suggests is "reopened" by Hut-ton's *Theory of the Earth* in 1795 (118). As Rossi points out, the French Revolution reactivated the moral sensitivity to materialism. Materialism also acquires a heightened charge in the wake of the Industrial Revolution, especially clear in the animated matter of a rock record narrativized at the instigation of the coal industry. Rossi shows that the his-tory of the earth continues to rely on the history of nations as a fictive organizing structure through the nineteenth century—it is fully absorbed into the truth-claims of geology only after two hundred years (120, cp. 45). In this way, he confirms that "so-called scientific progress often consists not so much in a progress within science as it does in taking some-thing that formerly was not science and making it part of science itself" (xiv).

32. *The Prose Works of William Wordsworth*, ed. W. J. B. Owen and J. Smyser (Oxford: Clarendon, 1973), 2:181. Theresa Kelley's fine analysis of this passage in *Wordsworth's Re-visionary Aesthetics* (Cambridge: Cambridge UP, 1988), 17–19, illuminates the aesthetic underpinnings of Wordsworth's geomorphology. I would urge, however, that the rock record as it is understood here is not inscribed by aesthetic imperatives, but rather provides a geo-logical, material basis for "the purposes of pleasure."

33. *The Supplement of Reading: Figures of Understanding in Romantic Theory and Prac-tice* (Ithaca: Cornell UP, 1990), 302.

CHAPTER THREE

"Great Frosts and . . . Some Very Hot Summers"

Strange Weather, the Last Letters, and the Last Days in Gilbert White's The Natural History of Selborne

STUART PETERFREUND

1. Gilbert White (1720–1793)

Gilbert White, the Oxford-educated English clergyman and naturalist whose *Natural History of Selborne* (1788; 1789) was the most popular work of natural history before Darwin's *Origin of Species* (1859), achieved that popularity by writing of the flora and fauna of his environs with familiarity, knowledge, and accuracy. He had a sense of the ecological wholeness and interactivity of the world, and he expressed that sense through careful observation and rendering the results of that observation in fine detail. White's editor, Richard Mabey, discusses White's 1774 epistolary account (in Letter 21, dated September 28, 1774, to Daines Barrington) of the swifts that he observed from "his house in the High Street, just across the square from the church where he was curate." According to Mabey, White possesses a rare gift: "His swifts are real, not puppets—'bundles of responses'—or links in some taxonomic chain. They are living birds in a living and closely observed situation. With no more than a hint of anthropomorphism, White suggests that their lives have a richness and rhythm of their own. And by describing them on a summer evening in a particular English village, he is able to awake in us the possibility of a human response to them."[1]

White is able to strike a rare balance. While his birds are living subjects and not merely Cartesian automata, and while they are not merely the stuff of an exercise in avian taxonomic classification of the sort that had been going on in

England from at least the time of John Ray, some of whose work is discussed below, White is able to balance the desire to render them from the life with the imperative to render them accurately. In Letter 16 to Barrington, in which White describes the nesting behavior of martins, "the exactness of his observations and of his language are inseparable. The martins, for example, do not just use 'grass,' but 'little bits of broken straws,' an image that is at once more sympathetically evocative of the energetic activities of these little 'martlets,' and also more helpful in understanding how they actually put their nests together" (*Selborne* xix–xx).

But despite a commitment to writing natural history based on close observation and rendered with the minutest accuracy possible, White, a clergyman-naturalist like his predecessors, wrote from within the same theodical framework as Ray and William Derham—a framework that had previously yoked natural philosophy, natural history, and physico-theology and which in White's time continued to yoke natural history and physico-theology under the aegis of natural theology.[2] Moreover, White wrote, as he understood his situation, from within the six thousand–year world chronology of Bishop Ussher. And that chronology was nearing its end as White wrote.

2. Strange Statistics, Stranger Analysis

The Natural History of Selborne bears a completion date of June 25, 1787 (*Selborne* 268). White's fifth letter to Thomas Pennant was written very close to that date—and certainly no earlier than January 1, 1787, the terminal date of the enclosed table that records local rainfall for the period May 1, 1779 to January 1, 1787. In discussing this particular table, White seems uneasy with the variability of the rainfall. He at once protests that he is "not qualified to give the mean quantity" over time, by reason of not having "measured it for a very long period" (16), yet he annotates with exclamation marks the entry for May 1, 1779 to January 1, 1780, which records 28.37 inches of rain in eight months, and the entry for January 1, 1782 to January 1, 1783, which records 50.26 inches of rain in a year. Further on, in his eighth letter to Barrington, dated December 20, 1770, White returns to the theme of excessive rainfall, remarking in a postscript that "There fell in the county of Rutland, in three weeks of this present very wet weather, seven inches and an half of rain, which is more than has fallen in any three weeks for these thirty years past in that part of the world. A mean quantity in that county for one year is twenty inches and an half" (130).

With the exception of the rainfall reported to Barrington, which would prorate to an annual rainfall total of some 130 inches, nearly 650 percent above the annual mean, these are not wild variations. Both entries annotated with exclamation marks fall within 50 percent of the mean of 35.91 inches per year for the period covered: 50.26 inches of rain is 40 percent above the annual mean; 28.37 inches of rain in eight months, prorated as an annual rainfall of 42.56 inches, is 18.5 percent above the annual mean. Moreover, White takes no such notice of the entry for January 1, 1780 to January 1, 1781, during which 27.32 inches of rain—24 percent *below* the annual mean—fell.

Furthermore, when White turns his attention to the twenty-year tables for baptisms, burials, and marriages from 1761 to 1780, he annotates no line in any of the columns with an exclamation mark, despite the occurrence of potentially ominous variations from the mean. Such variations include only thirteen baptisms in both 1768 and 1773, 32.6 percent below the twenty-year mean of 19.3; twenty-one burials in 1775, 70.7 percent above the twenty-year mean of 12.3; and only one marriage in 1774, 75.9 percent below the twenty-year mean of 4.15. Indeed, far from pronouncing himself incompetent to talk about the population, White notes proudly that it has grown by some 176, from "about 500" to 676, since the time that his grandfather and namesake was vicar of Selborne.

3. Selborne as Providential Paradise?

In his nonnumerical description of Selborne, White presents the village and its growing population as innocent, substantial, and unchanging, if not wholly Edenic. "We abound with poor, many of whom are sober and industrious, and live comfortably in good stone or brick cottages which are glazed, and have chambers above stairs: mud buildings have we none," White informs Pennant. Further on, White adds, "The inhabitants enjoy a good share of health and longevity; and the parish swarms with children" (*Selborne* 17–18). In this accumulation of detail, one can almost catch an echo of God's valedictory blessing of the newly created human inhabitants of the earth in the *P* account of the creation: "Be fruitful, and multiply, and replenish the earth, and subdue it, and have dominion over . . . every living thing that moveth upon the earth" (Gen. 1:28). Indeed, there is much in White's *Selborne,* here and elsewhere, to validate David Elliston Allen's assessment of this text. Citing James Russell Lowell's *My Study Windows* (1871) with approbation, Allen observes that "Lowell came very close to the truth in calling *Selborne* 'the journal of Adam in Paradise.' For it is, surely, the testament of Static Man: at peace with the world and with himself, content with deepening his knowledge of his one small corner of the earth, a being suspended in perfect mental balance. Selborne is the secret, private parish inside each one of us."[3]

Why should White's initial concern with the extremes of climate be exclusively with variations of rainfall above the mean? Why does he omit mention of and commentary on other kinds of strange weather? One obvious answer to the question is that the Bible identifies the flooding caused by forty days of rain in the time of

Noah as the cause that obliterated Eden as well as all other originary sites of creation. Genesis reports that "the waters prevailed exceedingly upon the earth; and all the high hills, that were under the whole heaven, were covered" (7:19).

But to begin to answer this question in greater detail, one ought to consider the way that two of White's important predecessors, the natural theologians John Ray and William Derham, treat the meaning of floods and rain. The logic of such a comparison seems obvious enough, given the fact that White cites Ray's *The Wisdom of God Manifested in the Works of the Creation* (1691; 7th ed., 1717) and Derham's *Physico-Theology; or, a Demonstration of the Being and Attributes of God from His Works of Creation* (1713; 4th ed., 1716), throughout his text, usually with approval.[4]

In a sense, White's initial focus on rainfall is overdetermined—not only by his theological commitment but by his commitment to the tradition of natural theology, of which Ray and Derham are exemplars, as well as important predecessors. For both Ray and Derham, meteorological phenomena consist of wind and rain and virtually nothing else—witness Ray's decision to limit his discussion to these two "Meteors," that is, meteorological phenomena, and Derham's decision to follow the lead of Ray, among others.[5] Though there is hot weather, there is virtually no drought, hard frost, snow, or hail in either's world, or text. Wind and rain in particular are signs of God's providential wisdom in working out a plan for this world, which Ray and Derham hold to be essentially unchanged and unchanging virtually from the time of the creation, and for the greater good of those in this life.

Ray, for example, after correctly identifying rain as the cooled, hence condensed, distillate of warm water vapor arising from the earth's large bodies of water, notes that the clouds from which the rain falls are "so carried about by the Winds, as to be almost equally dispers'd and distributed, no Part of the Earth wanting convenient Showers, unless when it pleaseth God, for the Punishment of a Nation, to with-hold Rain by a special Interposition of his Providence." He freely grants that there are exceptions, such "as the Land of Egypt, [where] though there seldom falls any Rain there, yet hath abundant Recompence made it by the Overflowing of the [Nile] River. This Distribution of Clouds and Rain is to me (I say) a great Argument of Providence and Divine Disposition."[6] Elsewhere, treating "of Fire, Hail, Snow, and other Elements and Meteors," among other things, Ray argues, following the precedent of Psalm 19:1 ("The Heavens declare the Glory of God . . ."), that "Man is commanded to observe and take notice of their curious Structure, Ends, and Uses, and give God the Praise of his Wisdom, and other attributes therein manifested" (*Wisdom* 178–79).

For Ray, such "Providence and Divine Disposition" have been in effect since the Old Testament God of Genesis made a covenant with Noah and his descendants that "the waters shall no more become a flood to destroy all flesh" (9:15). Otherwise, there might be in some Lands continual successive Droughts for many Years, till they were quite depopulated; in others, as lasting Rains, till they were overflown and drown'd . . . whereas since the ancientest Records of History we do not read or hear of any such Droughts or Inundations, unless perhaps that of *Cyprus,* wherein there fell no Rain there for Thirty six Years, till the Island was almost quite deserted. (*Wisdom* 88–89)

Indeed, the very manner in which rain falls bespeaks providential design. Considering "the Manner of the Rain's Descent," Ray remarks the way in which it falls "gradually, and by Drops, which is most convenient for the watering of the Earth; whereas, if it should fall down in a continual Stream, like a River, it would gall the Ground, wash away Plants by the Roots, overthrow Houses, and greatly incommode, if not suffocate Animals." No second deluge: hence, "If . . . we consider these Things, and many more that might be added, we might in this Respect also cry out with the Apostle, *O the Depth of the Riches both of the Wisdom and Knowledge of God!*" (89–90).

Even when it causes local flooding, the rain, according to Ray, accomplishes providential ends. In his analysis,

the Rain brings down from the Mountains and higher Grounds a great Quantity of Earth, and in Times of Floods spreads it upon the Meadows and Levels, rendering them thereby so fruitful as to stand in Need of no Culture or Manuring. So we see the Land of *Egypt* owes its great Fertility to the Annual Overflowing of the River *Nilus:* And it's likely the Countries bordering on the River of *Ganges* may receive the like Benefit by the Overflowing thereof. (82–83)

Derham, like Ray, limits his discussion to "two . . . Meteors, the Winds, and the Clouds and Rain,"[7] focusing in each instance on the usefulness of the meteorological phenomenon in question and arguing that such usefulness testifies to the providential design of the universe. The clouds, for example, provide "refreshing pleasant Shades," as well as

The fertile Dews and Showers which they pour down on the Trees and Plants, Which would languish and die with perpetual Drought, but are hereby made Verdant and Flourishing, Gay and Ornamental, so that (as the Psalmist saith, Psal. 65. 12, 13) *The little Hills rejoyce on every side, and the Valleys shout for joy, they also sing.*

And, if to these Uses, we should add the Origine of Fountains and Rivers, to Vapours and the Rains, as some eminent modern Philosophers have done, we should have another instance of *the* great Use and Benefit of that Meteor. (20–25)

By way of contrast, when White does talk about bad weather, which his preoccupation with above-average rainfall initially foregrounds as a cause for concern, his commitment to rendering the full range of observed natural phenomena prevents him from restricting his remarks to matters of wind and rain. Furthermore, the discourse manifests the element of catastrophism suggested by the exclamation marks, an element all but absent from the discourse of Ray and Derham. Not that White is particularly eager to broach the subject of meteorological catastrophes: it is only at the very end of his *Selborne,* after he has successfully repressed any consideration of such disastrous weather throughout the rest of his text, save for his report of flooding to Barrington and a brief mention of the frost of 1768 to Pennant, that White turns to consider "great frosts and, . . . some very hot summers" (*Selborne* 253).[8] The last six letters of White's *Selborne*—letters 61–66 to Barrington—detail meteorological catastrophes of one sort or another that occurred throughout virtually the full chronological sweep of the text, dating from January 1768 to June 1784. The first four of these last six letters deal with extremes of heat and cold; the last two deal with even more ominous phenomena such as meteors and thunderstorms.

Along with flooding, these are clearly meteorological catastrophes that White is mindful of in letter 5 to Pennant—the last occurred in 1784, some three years before the drafting of letter 5. In actual point of fact, as Anthony Rye observes, White quite probably "hated winters, possibly feared them, as all country dwellers then must [*sic*]"—this much "is clear from his Journal records kept for near thirty years. During those years some terrible winters came, and many formidable springs and summers too, of harsh winds, cold and hot droughts, prolonged rains which 'glutted' fields and pastures."[9] Nevertheless, White remained silent on such matters for between three and nineteen years, repressing any mention of these meteorological catastrophes through thirty-nine additional letters to Pennant and sixty to Barrington, even though at least two of the catastrophes that White finally does discuss occurred in their entirety several years before the time of the last letter to Barrington that is headed with a date (Letter 55, dated October 10, 1781 [*Selborne* 241]), and a significant part of one of the catastrophes discussed occurred several months before that time.[10]

4. Providence and Paradise in Conflict with a Changing World

White's repression arose from the fact that, as much as he would have liked it, he did not have the luxury—as Ray and Derham apparently did, to judge from their rhetoric—of seeing the world as unchanged and unchanging. His may have been "the journal of Adam in Paradise," but White's *Selborne* at once represses and recognizes the fact that any re-creation of Eden entails re-creating both the conditions that occasion humanity's loss of innocence and the consequent expulsion from Paradise. This preoccupation is evident from White's choice of English precursor texts to provide literary counterpoint and ornament throughout his text—he repeatedly cites Milton's *Paradise Lost,* as well as other texts by Milton.[11] These texts were as readily available in the time of Ray and Derham as they were in White's time, but they are as notable for their absence in the texts of these two predecessor–naturalists as they are for their presence in White's text.

There is an uncanny but hardly coincidental isomorphism between the weather at the margins of *Paradise Lost* and the weather at the margins of White's *Selborne*. The former text concludes with Adam and Eve paying the penalty for their freely willed, transgressive loss of innocence. The penalty includes, among other things, expulsion from Eden into a landscape the heat of which evokes a reminiscence of the fall of Satan and his minions into a realm of "Adamantine Chains and penal Fire."[12] The two descend from the Edenic mount, escorted by "the hastning Angel" through "th'Eastern Gate," where

> The brandisht Sword of God before them blaz'd
> Fierce as a Comet; which with torrid heat,
> And vapor as the Libyan air adust,
> Began to parch the temperate Clime. (12:633–39)

The first two of White's last three letters also deal with hot weather—as advertised, with "some very hot summers"—specifically, with the summers of 1781 and 1783, especially the latter. Both summers, according to White, "were unusually hot and dry" (*Selborne* 263). The latter "was an amazing and portentous one, and full of horrible phaenomena" (265).

As is the case with *Paradise Lost*, the hot weather marks a portentous falling off, if not the Fall itself. During the summer of 1781, the British army was being engaged regularly by the Continental army in the course of the southern campaign that would culminate in Cornwallis's surrender at Yorktown on October 19.[13] Much of that army was composed of Englishmen—especially those from the rural, agrarian areas of England—dispossessed and constrained to go abroad because of the depredations wrought by the capitalist greed that drove both the industrialization and enclosure of England.[14] Such greed, though hardly original, was the besetting sin of White's time. And during the summer of 1783, American and British delegates were putting the final touches on the Peace of Paris which, when signed on September 3, 1783, finally brought hostilities between England and its former colonies to closure with the recognition of the colonies' right to exist as a sovereign and independent nation. Given the fact that the New World was often figured as a new Eden, White's reports of such weather were certainly fortuitous, if not fortunate.[15]

The tension between the desire to see the world as unchanged and unchanging, and the data—especially meteorological data as a metonym for historical process—that suggest that the case is otherwise, is evident from White's first nine letters to Pennant. These letters, most probably composed and assembled by White "just before his manuscript went to press," comprise, as White's editor notes, "some kind of introduction to describe the scene in which the book is set" (*Selborne* 269 n.). In the very fictive space where he is engaged in presenting the setting, geography, and geology of Selborne in such a manner as to make these features appear unchanging, White also expresses concern over variations in rainfall that threaten his representations of stasis.

Perhaps the finest expression of this tension is to be found in letter 11, which describes "the centre of the village" in terms fraught with Edenic symbolism. In Genesis, "the tree of life . . . and the tree, of knowledge of good and evil" (2:9)[16] responsible for the Fall are both planted in the center of the garden. Similarly, at the center of White's Selborne are to be found the church, which as the symbol of everlasting life tropes "the tree of life," and "a square piece of ground surrounded by houses, and vulgarly called the Plestor." This square has a rather interesting history. "In the midst of this spot stood, in old times, a vast oak, with a short squat body, and huge horizontal arms extending almost to the extremity of the area." Along with its symbolization of druidic nature worship and English empire—English oak was the wood of choice for building the hulls and decks of the British navy and merchant fleet that conquered and colonized much of Africa, Asia, and North America—such a tree tropes "the tree of knowledge of good and evil." Perhaps more to the point, this tree, like the tree it tropes, participates in a fall of sorts, this one closely associated with violent weather: "Long might it have stood, had not the amazing tempest of 1703 overturned it at once, to the infinite regret of the inhabitants, and the vicar, who bestowed several pounds in setting it in its place again; but all his care could not avail; the tree sprouted for a time, then withered and died" (*Selborne* 10).

"[T]he amazing tempest of 1703" and its aftermath comprise an apt symbolization in this instance for events that threatened English military and mercantile supremacy of the seas in that time and throughout the last quarter of the eighteenth century. Occurring shortly after the accession of Queen Anne to the throne (reigned 1702–1714), the storm in question destroyed the symbol of that supremacy at the very time that it was being tested on the high seas by the French, who were launching raids against the English coast, much as they would do after the English declared war against the Jacobin government.[17] These raids by the Jacobins were but another point along a path of decline, following as they did England's defeat in the American Revolution. That defeat, caused in its turn by American naval victories over a British fleet of greater tonnage and superior firepower, resulted in the final, irrevocable loss of sizable colonial holdings and the income that these had produced, fit retribution, perhaps, for the policies that had populated those colonies with unwilling English expatriates in the first place.

White's discussion of the meteorological and natural–historical events of 1703 provides a way of understanding his *Selborne* as an attempt to mediate the tension between his desire to see the locale of Selborne (and of England more generally) as the type of an unchanging earthly Paradise, and the imperative to take account of and come to terms with change. Moreover, that discussion suggests that any such attempt at mediation must take account of a wide range of temporal change, ranging from the natural to the sociopolitical, up to and including the eschatological. Finally, White's discussion reveals a symbolic register in terms of which meteorological catastrophes stand for the manifestations of sociopolitical change threatening the established order of England.[18]

5. Stormy Weather

When White does discuss meteorological catastrophes with his correspondent Barrington, he begins by waxing unaccountably proleptic, starting off letter 16 with the following anticipation and preemption of any and all objections: "Since the weather of a district is undoubtedly part of its natural history, I shall make no further apology for the . . . following letters, which will contain many particulars concerning some of the great frosts and a few respecting some very hot summers, that have distinguished themselves from the rest during the course of my observations" (*Selborne* 253).

The first of the "great frosts" that White discusses is "the frost in January 1768 [which] was, for the small time it lasted, the most severe that we have known for many years" (253). What made this frost particularly destructive, according to White, was its snow, followed by the alternations of severe nighttime freezes with daytime thaws, "so that the laurustines, bays, laurels, and arbutuses looked, in three or four days, as if they had been burnt in a fire" (253). Ultimately, "it appeared [that] the ilexes were much injured, the cypresses were half destroyed, the arbutuses lingered on, but never recovered; and the bays, laurustines, and laurels, were killed to the ground; and the very wild hollies, in hot aspects, were so much affected that they cast all their leaves" (256). Nor was the destruction confined to plant life. "The coincidents attending this short but intense frost were, that the horses fell sick with an epidemic distemper, which injured the winds of many, and killed some; that colds and coughs were general among the human species . . . that several redwings and thrushes were killed by the frost" (254–55).

Not all the flora and fauna were so severely affected: "To the great credit of the Portugal laurels and American junipers, be it remembered that they remained untouched during the general havoc." So, too, "the large titmouse continued to pull straw lengthwise from the eaves of thatched houses and barns in a most adroit manner, for a purpose that has been explained already" (255).[19]

There are intimations of apocalypse here. The idea that snow, followed by the alternation of freeze and thaw, made the leafy evergreens indigenous to Selborne appear, in White's words, "as if they had been burnt in a fire" has some striking affinities with what John of Patmos reports happening when the first of "the seven angels which stood before God" sounds its trumpet "and there followed hail and fire mingled with blood, and they were cast upon the earth: and the third part of the trees was burnt up, and all green grass was burnt up" (Rev. 8:1, 7).

Another set of affinities arising from the specter of leafy evergreens looking "as if they had been burnt in a fire"—a set of affinities that is reinforced by the ever-increasing presence of the motif of fire and of the associated motifs of sulfurous exhalations and atmospheric portents throughout these last six letters—is with Thomas Burnet's *Sacred Theory of the Earth* (1684–1689; 1690–1691). Burnet's treatise undertakes to reconstruct, by means of some geological data and a good deal of biblical hermeneutics, the way that the flood occurred and the way that the con-flagration marking the end of the world is supposed to occur. And occur it shall, according to Burnet: "[T]he truth and certainty of the Conflagration" are unar-guable. The true object of inquiry is "the *Time, Causes,* and *Manner* of it."[20] Committed, as an Anglican priest, to Bishop Ussher's chronology, the same chronol-ogy to which Burnet is committed, White harbors a keen sense that theodical time is running out, and that meteorological and other signs portend just such an end.

Interestingly, the two flora that survive this apocalyptic frost the best are the "Portugal laurels and American junipers." "The large titmouse" (*Selborne* 255), if not specifiably British, is certainly indigenous to the British isles. Nor are the iden-tities of these life-forms necessarily coincidental.[21] The frost of which White writes occurred virtually in the midst of the events leading to the American Revolution (1776–1783), and the identities of these two of the three hardy survivors may be seen to symbolize the indomitable, if contrary, wills of the American colonies and England to survive and prevail in the inhospitable sociopolitical climate of the 1760s. Granted, White is writing from the perspective of 1787, in the aftermath of the American Revolution and the Peace of Paris (1783). But even at the begin-ning of 1768, an aura of apocalyptic political presentiment pervaded relations between England and its American colonies.

The Stamp Act, which precipitated the events leading to the revolution, had been passed by both houses of Parliament early in 1765 and was approved by George III on March 22 of that year. Although repealed a year later by the Pitt gov-ernment in March 1766, the Stamp Act was replaced by the Declaratory Act, reaffirming parliamentary authority over the colonies "in all cases whatsoever" at that time. A year and two months after passage of the Declaratory Act, in May 1767, the English government suspended the New York Legislature for failing to provide quarters for British soldiers in compliance with the Mutiny Act of 1765. A month later, in June 1767, Parliament passed the Townshend Acts, which placed heavy import duties on British lead, glass, paint, paper, and tea.

White's frost occurred in this niche of time, just prior to the Massachusetts As-sembly's decision on February 11, 1768 to put the Assembly's view of the problems at hand in the form of a circular letter to be forwarded to other colonial legislatures for comment and response. Although partially repealed in 1770, the Townshend Acts reappeared in a different guise as the Tea Act of 1773, which not only retained the impost on tea but also granted to the East India Company "a monopoly of the tea trade to the colonies." The Boston Tea Party (December 16, 1773) was but one response. Shortly thereafter, in early 1774, Parliament passed the Coercive Acts. With the passage of the Quebec Act in June 1774, which established "non-representative government, trial without jury, and the 'free exercise of the religion of the Church of Rome' in a greatly expanded Quebec,"[22] American colonists were alarmed at their diminished prospects for local autonomy. Lexington and Concord were ten months off; the Declaration of Independence, slightly more than two years away.

The prolepsis of letter 61 softens to litotes in letter 62, which begins: "There were some circumstances attending the remarkable frost in January 1776 so singular and striking, that a short detail of them may not prove unacceptable" (*Selborne* 256). But this softening does not obscure the fact that White has, by the time of writing this letter, repressed the discussion of this frost in 1776 for some ten or eleven years in the epistolary chronology of *Selborne*, as is indicated by his frank admission that "the most certain way to be exact will be to copy passages from my journal, which were taken from time to time as things occurred" (256).

Occurring more than eight months after the Battle of Lexington and Concord (April 19, 1775) and less than six months before the Declaration of Independence, this second frost, although arguably less apocalyptic than the first, is no less frightful for its disruptions of natural and human order. White verges on the diluvian, if not the apocalyptic, when he talks of how "Snow driving all day [January 7] . . . was followed by frost, sleet, and some snow, till the 12th, when a prodigious mass overwhelmed all the works of men, drifting over the tops of the gates and filling the hollow lane" (257).[23] And he likens the scene to the sort one might expect to encounter at the very fringes of civilization, referring to "Siberian weather" (*Selborne* 257; *Journals* 118) and to "a sort of Laplandian scene, very wild and grotesque indeed" (258). In a world turned upside down, usual and accustomed distinctions having to do with nature and culture, clean and unclean, are utterly and frighteningly subverted in a cacophonic (and cacophagic) feeding frenzy.[24] "Tamed by the season, skylarks settled in the streets of towns, because they saw the ground was bare; rooks frequented dunghills close to houses; and crows watched horses as they passed, and greedily devoured what dropped from them; hares now came into men's gardens, and scraping away at the snow, devoured such plants as they could find" (*Selborne* 258; *Journals* 118–19).[25]

No less frightening is the subversion of England-as-usual and the implied threat to the English way of life symbolized by freedom of movement in matters of state and commerce and the sounds announcing such movement. "The company at Bath, that wanted to attend the Queen's birthday, were strangely incommoded: many carriages of persons, who got, in their way to town from Bath, as far as Marlborough, after strange embarrassments, here met with a *ne plus ultra* (*Selborne* 257).[26] Neither money nor the class structure that money helps to reify avails against the weather, with the result that the nobility is reduced to the condition of common travelers. "The ladies fretted, and offered large rewards to labourers, if they would shovel them a track to London; but the relentless heaps of snow were too bulky to be removed; and so the 18th passed over, leaving the company in very uncomfortable circumstances at the Castle and other inns" (257–58).[27]

Four days later, White himself had occasion to go to London, which exhibited the sort of subversion of categories he had encountered among the wild creatures in the countryside. Being bedded deep in snow, the pavement of the streets could not be touched by the wheels or the horses' feet, so that the carriages ran about without the least noise. Such an exception from din and clatter was strange, but not pleasant; it seemed to convey an uncomfortable air of desolation: "'ipsa silentia terrent' [a silence terrifying in itself]" (258).

In summarizing the toll taken by this frost on the flora and fauna in the environs of Selborne, White notes that the former came through in relatively good shape: "No evergreens were quite destroyed; and not half the damage sustained that befell in January, 1768." The latter were not so fortunate, however. White remarks a decline in the songbird population, and with that decline a movement toward the silence that had seemed so terrifying in London. And he makes a point of linking at least a portion of the destruction of the bird population to the concurrent falling off of English society signaled by poaching, one of the illicit means that rural laborers had of obtaining food after being driven from the land by enclosure. "As to the birds, the thrushes and blackbirds were mostly destroyed; and the partridges, by the weather and poachers, were so thinned that few remained to breed the following year" (260).

Litotes prevails in letter 63 to Barrington, who, White "trust[s], will not be displeased to hear the particulars" (260) of the frost of December 1784. Although probably more destructive than any other frost on record, including that of the winter of 1739–1740 (261 n.), this third frost coming after the other two, as well as more than one year after the Peace of Paris, combines the apocalyptic with the familiar and the domesticated. As to the damage it caused to his leafy evergreens, White laments, "all my laurustines, bays, ilexes, arbutuses, cypresses, and even my Portugal laurels, and . . . my fine sloping laurel hedge were scorched up" (261).[28] Subsequently, he reports, "This frost killed all the furze and most of the ivy, and in many places stripped the hollies of their leaves. It came at a very early time of the year, before old November ended" (262). The loss of these last two plants in early December no doubt had an impact on the Christmas season to follow, holly and ivy being the stuff of English carols and holiday decorations alike.

As to hail-like manifestations from on high, White observes "that on Friday, December the 10th, being bright sunshine, the air was full of icy spiculae, floating in all directions, like atoms in a sunbeam let into a dark room. . . . Were they watery particles of the air frozen as they floated; or were they evaporations from the snow frozen as they mounted?" (262). White leaves the question unanswered. One other phenomenon of possibly apocalyptic import is the static electricity that results from the ionization of the air brought on by the cold: "during those two Siberian days [of December 9–10], my parlour-cat was so electric, that had a person stroked her, and been properly insulated, the shock might have been given to a whole circle of people" (262).

A shock of this sort might well symbolize the breaking of the circle—the disruption of social relations, of England-as-usual. Indeed, in the very next paragraph, White realizes, "I had forgot to mention that, during the two severe days, two men, who were tracing hares in the snow, had their feet frozen, and two men, who were much better employed, had their fingers so affected by the frost, while they were thrashing in a barn, that a mortification followed, from which they did not recover for many weeks" (262). Idle agrarian laborers and the sturdy poor alike contract frostbite and suffer a common fate of being stymied in their pursuits in weather of the sort that White describes.

In turning from his discussion of the three frosts to that of the three hot summers in letter 64, White evinces the sense that, by going on as he has, he may have sensationalized the weather in question more than he intended to, and in so doing may have revealed more of his personal anxieties than he wished to. Accordingly, he begins by stating that since "the effects of heat are seldom very remarkable in the northerly climate of England . . . I shall be more concise in my account of the severity of a summer season, and so make a little amends for the prolix account of the degrees of cold, and the inconveniences that we suffered from late rigorous winters" (263).[29]

The summer of 1781 was one in which another "third part of trees was burnt up" (Rev. 8:7). White reports, "my peach and nectarine trees suffered so much from the heat that the rind on the bodies was scalded and came off; since which the trees have been in a decaying state. . . . During that summer also, the apples were coddled, as it were, on the trees; so that they had no quickness of flavour, and would not keep the winter" (*Selborne* 263).

This summer was not without its omens as well. White, who was following the final stages of the Revolutionary War quite closely and seemed persuaded that the red-clad British forces under Cornwallis were destined to be the victors, reports suggestively on the red planet in the journal entry for July 21: "The planet Mars figures every evening & makes a golden & splendid show.[30] This planet being in opposition to the sun, is now near us, & consequently bright" (*Journals* 189). But there were contrary omens as well. The entry for June 30 reports that "a large meteor appeared falling from the S. toward the E. in a [*sic*] inclination of about 45 degrees, & parting in two before I lost sight of it." The entry for the very next day is simply "Wheel around the sun" (187). Although White reports, in the entries for July 14 and 15, a summer that is, near its outset, so wet that the "soft rains" spoil the hay harvest and cause "[t]he farmers [to] complain of smut in their wheat" (188–89), the excessive moisture ultimately gives way to drought. The entry for August 11 reads, "Ponds & streams fail People in many parts in great want of water. The reapers were never interrupted by rain one hour the harvest thro" (191).

These entries call to mind John of Patmos's account in *Revelation* of what happened after "the third angel sounded, and there fell a great star from heaven, burning as it were a lamp, and it fell upon the third part of the rivers, and upon the fountains of waters." This "star . . . called Wormwood" made that third of the waters lethally unfit to drink (8:10–11). Not surprisingly, in the very next verse, "the fourth angel sounded, and the third part of the sun was smitten, and the third part of the moon, and the third part of the stars; so as the third part of them was darkened, and the day shone not for a third part of it, and the night likewise" (8:12).[31]

The summer of 1783, by way of contrast, according to White, was one in which "myriads" of wasps "would have devoured all the produce of my garden, had we not set the boys to take the nests, and caught thousands with hazel twigs tipped with bird-lime" (*Selborne* 263). Here, England-as-usual, symbolized by agrarian labor prudently deployed, prevails over a menace that, even if it is accurately rendered, resembles the locusts of Exodus and Revelation a good deal.[32] But wasps were not the only plague of this summer. "In the sultry season of 1783, honey-dews were so frequent as to deface and destroy the beauties of my garden. My honeysuckles, which were one week the most sweet and lovely objects that the eye could behold, became the next the most loathsome; being enveloped in a viscous substance, and loaded with black aphides, or smother-flies" (264).[33]

The journal entries for June 23 and 24 of this summer are not without their apocalyptic and literary overtones. In the former entry, White mentions the honey-dew, then changes the subject and deploys the same sort of freeze–burn description he had used to characterize the hard frosts discussed above: "Vast honey-dew; hot & hazey; misty. The blades of wheat in several fields are turned yellow, & look as if scorched with the frost" (*Journals* 222). In the latter entry, White returns to the weather, characterizing the light conditions in terms of the heroic simile (*Paradise Lost*, 1:594–99), discussed below, that likens Satan's face to the just-risen sun: "Vast dew, sun, sultry, misty & hot. This is the weather that men think is injurious to hops, The sun 'shorn of his beams' appears thro' the haze like the full moon" (222).

The summer of which White writes was the one immediately preceding the final ratification of the Peace of Paris, and it is extremely tempting to read his account of this onset of plagues as being in some way tied to Britain's procrastination in giving the thirteen American colonies their final and unconditional freedom. White, by the way, would not have been the only writer ever to have seen parallels between Rameses II and George III as monarchs distinguished both for extraordinarily long reigns and for their efforts at the colonization and subjugation of other lands and peoples.[34]

More than plagues contributed to the "amazing and portentous" (*Selborne* 265) summer of 1783, which was full of harbingers of imminent apocalypse: "for besides the alarming meteors and tremendous thunder-storms that affrighted and distressed the different counties of this kingdom, the peculiar haze, or smokey fog, that prevailed for many weeks in this island, and in every part of Europe, and even beyond its limits, was a most extraordinary experience, unlike anything known within the memory of man" (265). Mabey notes that this atmospheric condition had a perfectly understandable explanation: "the eruption of the volcano Skaptárjökull, in Iceland—one of the most cataclysmic in historic times" (283 n.).

But this explanation was not widely known—White, although a man well versed in the natural sciences, does not seem aware of the cause of the phenomena he describes, for example—and the eruption, like that of Krakatoa almost exactly a century later, on August 26–27, 1883, caused widespread alarm and a tendency to think that the last days were at hand.[35] "The country people began to look with a superstitious awe at the red, louring aspect of the sun; and indeed there was reason for the most enlightened person to be apprehensive; for, all the while, Calabria and part of the isle of Sicily, were torn and convulsed with earthquakes; and about that juncture a volcano sprang out of the sea on the coast of Norway" (265), he reports.[36]

It is not difficult to imagine why these phenomena were considered harbingers of apocalypse, evoking as they do specific verses of Revelation. The general aspect of earthquake and smoke, for example, calls to mind the opening "of the bottomless pit; and there arose a smoke out of the pit, as the smoke of a great furnace; and the sun and the air were darkened by reason of the smoke of the pit" (9:2). And the "volcano [that] sprang out of the sea on the coast of Norway" calls to mind "a great mountain burning with fire cast into the sea: and a third part of the sea became blood" (8:8). But perhaps most compelling is "the red, louring aspect of the sun," which calls to mind the celestial manifestation of "a woman clothed with the sun, and the moon under her feet," and "a great red dragon, having seven heads and ten horns, and seven crowns upon the heads" (12:1, 3). The dragon symbolizes temporal power, and with its demise, John prophesies, will come the simultaneous end of the temporal power wielded by kings and priests, and human time.

Not long after these verses, John writes of "war in heaven: Michael and his angels fought against the dragon," also known as "that old serpent, called the Devil, and Satan, which deceiveth the whole world" (12:7, 9). These are clearly verses taken to heart by Milton in casting the "argument," or plot, of *Paradise Lost.* White also makes this association, concluding letter 65 with a heroic simile drawn from Milton's epic. White uses the simile to liken observed atmospheric conditions to those

> As when the Sun new ris'n
> Looks through the Horizontal misty Air
> Shorn of his Beams, or from behind the Moon
> In dim Eclips disastrous twilight sheds
> On half the Nations, and with fear of change
> Perplexes Monarchs. (1:594–99)

What White does not do is point to the object of the comparison: the face of Satan himself.

> Dark'n'd so, yet shone
> Above them all th' Arch-Angel: but his face
> Deep scars of Thunder had intrencht, and care
> Sat on his faded cheek, but under Brows
> Of dauntless courage, and considerate Pride
> Waiting revenge. (1:599–604)

To look at the skies of the summer of 1783 was to took at the face of Satan himself in Hell.

The apocalyptic associations continue in letter 66, the final letter. Perhaps because of the associations prompted by the object of the Miltonic simile on which he ended his previous letter, "his face" bearing "Deep scars of Thunder," White turns in this letter to thunderstorms, even though, as he candidly notes, the inhabitants of Selborne "are very seldom annoyed with thunder-storms," owing to the hills that intervene between the storms' usual origination point to the south and the village itself (266). However, "very seldom" does not mean "never." White recalls the storm of June 5, 1784 for his correspondent. White begins his account with meteorological data, but he soon moves away from it: "The thermometer in the morning being at 64, and at noon at 70, the barometer at 29, six-tenths, one half [29.65], and the wind north, I observed a blue mist, smelling strongly of sulphur, hanging along our sloping woods, and seeming to indicate that thunder was at hand" (266–67).

The gas in question is more likely ozone than sulfur, since large charges of electricity such as those packed by lightning act upon oxygen (O_2) to form its allotrope ozone (O_3) and since there is very little gaseous sulfur, whether hydrogen sulfide (H_2S) or hydrogen sulfite (H_2SO_3), in the atmosphere. But volcanic (or hellish) exhalations such as hydrogen sulfide (or brimstone) are clearly the more fitting accompaniment to the end of the world[37]—or, at the very least, the end of an English gentleman's dinner—than ozone.

> At about a quarter to two the storm began in the parish of Hartley, moving slowly from north to south. . . . We were just sitting down to dinner; but were soon diverted from our repast by the clattering of tiles and the jingling of glass.

There fell at the same time prodigious torrents of rain on the farms above-mentioned, which occasioned a flood as violent as it was sudden. . . . Those that saw the effect which the great hail had on ponds and pools say that the dashing of water made an extraordinary appearance, the froth and spray standing up in the air three feet above the surface. The rushing and roaring of the hail, as it approached, was tremendous. (267)

Much of what White includes in this letter to Barrington is taken verbatim from the journal entry of June 6 (*Journals* 242). The entry for the previous day is, if anything, more ominous and shows White even more alarmed than in the letter: "Much damage done to the corn, grass & hops by the hail; & many windows broken! Vast flood at Gracious street! vast flood at Kaker bridge [S. of Alton]. Hail near Norton two feet deep. Nipped-off all the rose-buds on the tree in the yard opposite the parlor window in order to make a bloom in the autumn. No bloom succeeded" (242).

What is there left to say? White closes by lamenting that he has not been able to write "an *Annus Historico-naturalis*, or the Natural History of the Twelve Months of the Year," which would have included "many incidents and occurrences that have not fallen in my way to be mentioned in my series of letters." But White commends to his correspondent Barrington the work of John Aikin—"Mr. Aikin of Warrington[, who] has lately published somewhat of this sort"—then "take[s] a respectful leave of you [i.e., Barrington] and natural history together" (*Selborne* 268).

Beneath the surface of this decorous leave-taking lies White's own horror in response to some of the phenomena he has described, as well as his fear that *Selborne* would meet with instant oblivion for its gloomy concluding pages. It is one thing to prefigure the end of the world of Grub Street, as Pope does in the *Dunciad*,[38] another entirely to prefigure the end of the world itself, as White does in *Selborne*.

White expresses his misgivings in "To Myself Commencing Author," a satire in the vein of Swift written at about the time of *Selborne*'s publication. In the satire, White likens his own fate as an author to that of another celebrant of nature (and of the Royal Society),[39] Abraham Cowley. In so doing, White expresses the fear that his literary remains, which he had once thought of as his "child," or posterity, may be no more than so much solid waste, to be offered jumbled together with other such waste to "Cloacina"—presumably, the presiding female deity of the sewer (L. *cloaca*).[40]

> Go, view that House amid the garden's bound,
> Where tattered volumes strew the teamed ground,
> Where Novels—Sermons in confusion lie,
> Law, ethics, physics, school-divinity;
> Yet did each author with a parent's joy
> Survey the growing beauties of his boy,

Upon his new-born babe did fondly look,
And deem Eternity should claim his book.
Taste ever shifts in half a score of years,
A changeful public may alarm thy fears;
Who now reads Cowley?—The sad doom await,
Since such as these are now may be thy fate,
For Cloacina with resistless sway
Demands her right and authors must obey.[41]

In the immediate aftermath of *Selborne*'s publication, it seemed as though White might be right about its reception and ultimate fate: "Superficially unattractive, a mere volume of letters, the book was slow in achieving popularity. . . . [It was only with the publication of Sir William Jardine's edition of 1827 that the real vogue for the book seems to have started and its literary merits to have become widely appreciated." David Allen freely owns that "it is difficult to account for this," speculating that perhaps "the style of writing was too Augustan to appeal immediately to a public accustomed to a diet of Romanticism."[42] But this speculation does little more than bandy about vague markers of periodization used by contemporary English departments, not eighteenth- and nineteenth-century readers.

Closer to the truth of the matter is the fact that in "the late 1820s . . . a new, no-nonsense, mainly middle-class audience emerged, able to discern the one great quality in White that has helped to bring his book such permanent renown: his gift of empathy, his ability to infuse deep feeling into what he described and recorded so carefully and soberly."[43] What needs to be added to Allen's observations to make them—and this discussion of *Selborne*—complete is that this new audience was sufficiently "no-nonsense" to disentangle the theology and its presumed authority from that "deep feeling." This new power of discernment came only after "the increasingly complex networks of discourse and, no less important, its ideological divisions . . . transformed the bible into a conflicted scene of interpretation."[44] Having been told in turn that any number of natural catastrophes, in conjunction with the American Revolution, the French Revolution, the rise of Napoleon, and the demise of Napoleon, presaged the end of the world, readers from the 1820s onward have been able to regard *Selborne* as something like a fabulated "journal of Adam in Paradise," treating of first things in a figurative, not a literal, sense, without brooding overmuch on how fabulated first things lead to historical last days.

Notes

1. Gilbert White, *The Natural History of Selborne*, ed. Richard Mabey (Baltimore: Penguin Books, 1977), ix–x. Subsequent citations of this work will appear parenthetically in the text and notes and will be to *Selborne*, followed by page number(s).

2. See Walter S. Scott, *White of Selborne* (London: Falcon, 1950), 223. In a letter to Robert Marsham that is quoted by Scott, White writes, "I concur with you heartily in your admiration of the harmony and beauty of the worlds of the creation. Physico-theology is a noble study, worthy of the attention of the wisest man."

3. David Elliston Allen, *The Naturalist in Britain: A Social History* (London: Allen Lane, 1976), 50.

4. See Scott, *White of Selborne* 218. In his review of *The Natural History of Selborne*, which appeared in the *Gentleman's Magazine* for 1789 and is quoted by Scott, Gilbert White's brother Thomas finds him "no unequal successor of Ray and Derham." Ray tends to be the authority of record regarding birds; Derham, the authority of record regarding insects. Speaking of Ray, for example, in the forty-third letter to Pennant (undated but probably written sometime between March 9, 1775 and November 30, 1780), White remarks on collecting a female specimen of *buteo apivorus, sive vespivorus Raii:* "The hen-bird was shot, and answered exactly to Mr. Ray's description of that species; had a black cere, short thick legs, and a long tail" (*Selborne* 101). And in the seventh letter to Barrington, dated October 8, 1770, White extends Ray's observation "that birds of the gallinae order, as cocks and hens, partridges, and pheasants, etc., are pulveratrices, such as dust themselves, using that method of cleaning their feathers, and ridding themselves of their vermin" (125).

Of Derham, White notes in the thirty-third letter to Pennant, dated March 30, 1771, that "There is an *oestrus*, known in these parts to every ploughboy; which, because passed over by Linnaeus, is also passed over by late writers, and that is the *curvicauda* of old Mouffet, mentioned by Derham in his *Physico-theology*, 250" (85). And in the fifty-third letter to Barrington, undated but probably written between September 3 and October 10, 1781, White, discussing insect migration, adds the following cite: "For various methods by which insects shift their quarters, see Derham's *Physico-Theology*" (238 n.).

Neither Thomas Burnet, Ray and Derham's contemporary, nor Burnet's *The Sacred Theory of the Earth* (1684–1689; 1690–1691) is anywhere mentioned—and for good reason. Burnet's account of *the* last days comports in several important ways with White's account of apocalyptic weather at the end of *Selborne*. Burnet will be discussed below.

5. See John Ray, *The Wisdom of God Manifested in the Works of the Creation*, 7th ed. (1717; New York: Arno, 1977), 88–91; and William Derham, *Physico-Theology; or, a Demonstration of the Being and Attributes of God from the Works of His Creation*, 4th ed. (1716; New York: Arno, 1977), A5v, 15–26.

6. Ray, *Wisdom*, 88–89.

7. Derham, *Physico-Theology*, 14.

8. The last paragraph of letter 13 to Pennant, dated January 22, 1768, does summarize the essentials of the frost of 1768, which occupies the whole of letter 61 to Barrington. But the letter to Pennant, while stating White's belief that "some days were more severe than any since the year 1739–40," nevertheless finds it "very providential that the air was still, and the ground well covered with snow, else vegetation in general must have suffered prodigiously" (*Selborne* 42). By way of contrast, letter 28 to Barrington, dated January 8, 1776, one day after the onset of the frost of 1776, omits any mention of it, as does letter 29, dated February 7, 1770, written less than a week after the thaw that ended this frost.

9. Anthony Rye, *Gilbert White and His Selborne* (London: William Kimber, 1970), 150–51.

10. Letter 61 recounts "the frost in January 1768" (*Selborne* 253); letter 62, "the remarkable frost in January 1776" (256). Letter 54 deals with "the summers of 1781 and 1783," which "were unusually hot and dry" (263).

11. For example, White cites *Paradise Lost* 2:476–77 (*Selborne* 104), 2:39 (112), 7:426–30 (132), 3:50 (175), 7:443–44 (218), and 1:594–99 (265–66), as well as *Il Penseroso* 81–82 (227). He also cites, among others, Shakespeare (118), John Philips's *Cyder* (221), and *Summer* from James Thomson's *Seasons* (27–28). White also shows an admirable command of classical texts, citing, among others, Oppian's *Cynara* (43); Aristotle's *History of the Animals* (43); Virgil's *Georgics* (66) and *Aeneid* (161, 194–95); and Lucretius's *De rerum natura* (206–07) and *Historia natura* (213). These are suggestive, not exhaustive, citation inventories.

12. *Paradise Lost*, 1:48, in *John Milton: Complete Poetry and Selected Prose*, ed. Merritt Y. Hughes (New York: Odyssey, 1957). Subsequent citations by book and line will appear parenthetically in the text.

13. See *Journals of Gilbert White*, ed. Walter Johnson (Cambridge, MA: MIT Press, 1971), 176, 184, 196. Subsequent citations of this work will appear parenthetically in the text and notes and will be to *Journals*, followed by page number(s). White apparently followed the final stages of the war quite closely. A journal entry for August 16, 1780, reports that "Lord Cornwallis gained a signal victory over General Gates in South Carolina, near Camden." A journal entry for February 15, 1781 reports that "On this day Lord Cornwallis gained a considerable victory over Gen. Greene at Guildford [*sic*] in N. Carolina." And a journal entry for October 19, 1781, reports that "On this ill-fated day Lord Cornwallis, & all his army Surrendered themselves prisoners of war to the united forces of France & America at York-town in Virginia."

14. See Laurence Goldstein, *Ruins and Empire: The Evolution of a Theme in Augustan and Romantic Literature* (Pittsburgh: U of Pittsburgh P, 1977), 104.

15. Walter Johnson, the editor of White's journals, warns that White is "erroneously thought to have concerned himself but little" with "the outside world." According to Johnson, White "was a strict constitutionalist, and feared outbreaks of violence, yet he trusted in the sturdy good sense of his countrymen" (*Journals* xxxii–xxxiii).

See also Rye, *White and His Selborne*, 155–57. Rye quotes White's political self-assessment, given somewhat later, after the outbreak of the French Revolution: "I was born a gentleman, and hope to remain one. I am no Jacobin" (157). Despite this self-assessment, however, White refrained from—indeed, actively opposed—at least one form of exploitation of the laboring classes by the upper classes: enclosure. Drawing on the record of White's successful opposition to "a bill for Selborne's enclosure [presented] to parliament," as well as testimonial evidence entered in the Selborne Parish Register in White's own hand, Rye characterizes him as "the watch-dog for the village." See also Raymond Williams, *The Country and the City* (London: Chatto and Windus, 1973), 229.

Goldstein, *Ruins and Empire* (104–05) suggests an ulterior motive for White's opposition to enclosure in his discussion of Oliver Goldsmith's portrayal of enclosure in *The Deserted Village*. Goldstein, eager to demonstrate that "trade's proud empire hastes to swift decay" (427), follows with particular attention those rural folk who of necessity volunteered for settlement or service in the new colonies. The "poisonous fields with rank luxuriance crowned" (351) shoot out from the "bloated mass of rank unwieldly woe," which Goldsmith in *The Traveller* described as luxury's ruinous impact on the kingdom. The infections are as universal as the empire; they reach into America as they do into Auburn and Grub Street.

From the perspective of the last six *Selborne* letters, then, England's defeat in the American Revolution may be seen as fit retribution for the policy of enclosure, which constituted a violation of England's "natural," rural agrarian order.

16. See also *Paradise Lost* 7:542–47.

17. Rye, *White and His Selborne*, 158, characterizes White's reaction to 1789 and its aftermath by observing, "The state of affairs in France was enough to disturb anybody. Why, it seemed we were suddenly at war, and raids were being made along the coast, just as in the time of Queen Anne! We nowadays are not unduly aware of there having been any raids on the south coast in Anne's reign, but so it had been, and now the twenty odd miles between Selborne and Portsmouth and Southampton had suddenly become quite small again."

18. See Leo Damrosch, *Fictions of Reality in the Age of Hume and Johnson* (Madison: U of Wisconsin P, 1989), 69–78. Damrosch notes the ways in which White bends natural history to buttress versions of the argument from design intended to defend the English establishment from its critics.

19. The purpose, explained in letter 41 to Pennant, undated but probably written between the dated letters of 2 September 1774 (letter 40) and 9 March 1775 (letter 42), is to feed on the flies concealed by the thatch (*Selborne* 98).

20. Thomas Burnet, *The Sacred Theory of the Earth* (1684–1689; 1690–1691; Carbondale: Southern Illinois UP, 1965), 253.

21. "Every species of titmouse winters with us" (*Selborne* 98). Less certain is whether "Portugal laurels" possess any symbolic significance. Portugal, having suffered the disastrous Lisbon earthquake in 1755, was definitely not a major power in the revolutionary politics of eighteenth-century Europe. If White's mention of this shrub has symbolic significance, it may be of the order of praise for Portuguese tenacity in the face of cataclysmic adversity. It is worth noting that although the laurel is mentioned twice thereafter (see 260, 261), the juniper is not.

22. Clinton Rossiter, *Seedtime of the Republic: The Origin of the American Tradition of Political Liberty* (New York: Harcourt Brace, 1953), 319–24.

23. "Fifteen cubits upward did the waters prevail; and the mountains were covered" (Gen. 17:20). "Every valley shall be exalted, and every mountain and hill shall be made low: and the crooked shall be made straight, and the rough places plain" (Isa. 40:4).

24. The mock apocalyptic vision of a jumbled or subverted English sociopolitical infrastructure was a recurrent literary motif throughout the eighteenth century, as glimpsed in works such as Swift's "Description of a City Shower" (1716), with its dark vision of a welter of the artifacts of urban living reduced to common sewage, or Pope's *Dunciad* (1741), with the community of letters reduced to universal darkness. Nearer in time to White's *Selborne*, the ditty "The World Turned Upside Down," which figured a topsy-turvy world, was as popular in its own way as "Lillabulero" was in its. When Cornwallis surrendered at Yorktown on October 19, 1781, it is the music he ordered the British army to play by way

of noting his view of the state of affairs—a mighty colonial power, its armed forces bested by the upstart inhabitants of the colonies in which those forces had been the representatives of British rule. Blake satirizes the specter of the sociopolitical topsy-turvy feared by a horrified establishment reacting to revolution in *The Marriage of Heaven and Hell* (1790–1793).

25. If anything, the excerpts from White's journals found in *Selborne* are not nearly so horrific as some of the others found in the journals themselves. "Hares, compelled by hunger, come into my garden & eat the pinks. Lambs fall, & are frozen to the ground. . . . As intense frost usually befalls in Jan: our Saxon fore-fathers call'd the month with no small propriety wolf-month; because the severe weather brought down those ravenous beasts out of the woods among the villages" (*Journals* 119). See also Scott, *White of Selborne*, 179; Rye, *White and His Selborne*, 174.

26. "The road-waggons are obliged to stop, & the stage-coaches are much embarrassed" (*Journals* 118).

27. The degree to which White views such weather as a threat to England-as-usual is suggested by his summary, "and so the 18th passed over," with its echo of Exodus and the narrative explaining how the Passover got its name.

> I will pass through the land of Egypt this night, and I will smite all the first-born in the land of Egypt, both man and beast; and against all the gods of Egypt will I execute judgment: I am the LORD. And the blood [on the doorposts] shall be to you for a token upon the houses where ye are: and when I see the blood, I will pass over you, and the plague shall not be upon you to destroy you, when I smite the land of Egypt" (Ex. 12:12–13).

28. White's journal entry for December 31 shows his awareness of the paradoxicality of the idea that frost burns: "My laurel-hedge, & laurustines, quite discoloured, & burnt as it were with the frost." That this observation may harbor apocalyptic overtones is suggested by the diluvian statistics that follow—in the words of White's editor Johnson, "[columns showing the monthly rainfall for 1784, at Fyfield, Selborne, & S. Lambeth respectively. Selborne has much the greatest total]" (*Journals* 254).

29. White is as good as his word: the first three letters cover ten pages in the Penguin edition; the last three cover five and one-half.

30. See note 13 above. On a related theme, a curious pair of journal entries is that of August 5 and 6. The former reports in part, "On this day a bloody & obstinate engagement happened between Admiral Hyde Parker, & a Dutch fleet off the Dogger-bank." The latter reports in part that "Every ant-hill is in a strange hurry & confusion; & all the winged ants . . . bent on emigration, swarm by myriads in the air, to the great emolument of the hirundines, which fare luxuriously. Those that escape the swallows return no more to their nests, but . . . lay the foundation for future colonies" (*Journals* 190–91).

31. See Burnet, *Sacred Theory*, 270. Here, White and his authority, John of Patmos, would appear to disagree with Burnet, for whom "the Conflagration had nothing to do with the Stars and superior heavens, but was wholly confin'd to this Sublunary World."

32. In Exodus, Moses warns Pharaoh, "if thou refuse to let my people go, behold, to morrow will I bring the locusts into thy coast: And they shall cover the face of the earth, that one cannot be able to see the earth: and they shall eat the residue of that which is es-

caped, which remaineth unto you from the hail, and shall eat every tree which groweth for you out in the field" (10:4–5). The locusts of Revelation, much like the vengeful Jehovah–God of the Passover on his mission against the firstborn, pursue sinful (unmarked) human beings rather than devouring their crops: "And it was commanded them that they should not hurt the grass of the earth, neither any green thing, neither any tree; but only those men who have not the seal of God in their foreheads" (9:4). These locusts inflict harm not with their mandibles but with their stingers: "And they had tails like unto scorpions, and there were stings in their tails: and their power was to hurt men five months" (9:10).

33. White's editor Mabey notes that "Honeydew is, in fact, a secretion produced by these aphids, and not the distilled essence of flowers" (*Selborne* 283 n.).

34. See Allen, *The Naturalist in Britain*, 43. Allen complicates the plot somewhat with the intriguing observation that the revival of interest in natural history in the mid-eighteenth century "was evident in a number of entirely separate quarters. One of them, extraordinarily, was at the topmost level in the land: the accession of George III in 1760 had among its many effects the minor and rather freakish one of placing at the nation's head, for the brief space of two and a half years, a trio of enthusiastic botanists." The trio comprised the Queen; Princess Augusta, the Queen's mother; and the Earl of Bute, George III's chief minister. Shelley's sonnet "Ozymandias" (1817) turns on the same implied comparison.

35. See Tom Zaniello, *Hopkins in The Age of Darwin* (Iowa City: U of Iowa P, 1987), 125. Discussing John Ruskin's "Storm-Cloud of the Nineteenth Century" (1883), Zaniello argues that Gerard Manley Hopkins's attempts to understand the atmospheric effects of the Krakatoa eruption are far more measured than "Ruskin's association of bad weather with a plaguelike vision of God's wrath."

36. See Burnet, *Sacred Theory*, 273. The references to earthquakes in Italy and the association of those convulsions with a volcano call to mind Burnet's discussion of Aetna and Vesuvius as harbingers of the final conflagration: "Let us therefore take these two Volcano's as a pattern for the rest; seeing they are well known, and stand in the heart of the Christian World, where, 'tis likely the last fire will make its final assault."

37. Burnet in the *Sacred Theory* argues:

> We may reasonably suppose, that, before the Conflagration, the air will be Surchargd every where . . . with hot and fiery exhalations; And as against the Deluge, whose regions were burthened with water and moist vapours, which were poured upon the Earth, not in gentle showres, but like rivers and cataracts from Heaven; so they will now be fill'd with hot fumes and sulphureous clouds, which will sometimes flow in streams and fiery impressions through the Air, sometimes make Thunder and Lightnings, and sometimes fall down upon the Earth in clouds of Fire. (277)

38. As Johnson points out, White had some sort of acquaintance with Pope: "When White took his bachelor's degree in 1743, no less a person than Alexander Pope presented him with his translation of the *Iliad* in six small volumes" (*Journals* xx).

39. White himself notes in *Selborne* that letters 16, 18, 20, and 21 to Barrington had previously been published in *Philosophical Transactions*, the journal of the Royal Society

(*Selborne* 145). Cowley, among other works, wrote the dedicatory ode for Sprat's *History of the Royal Society* (1660).

    40. See note 24 above.

    41. Quoted in Scott, *White of Selborne*, 211.

    42. Allen, *The Naturalist in Britain*, 50.

    43. Ibid., 50.

    44. Jon Klancher, *The Making of English Reading Audiences, 1790–1832* (Madison: U of Wisconsin P, 1987), 161.

Part II

The Global Reach of Natural History

CHAPTER FOUR

Jefferson's Thermometer

Colonial Biogeographical Constructions
of the Climate of America

ALAN BEWELL

[T]he Bard was weather-wise

—Samuel Taylor Coleridge

On the morning of July 4, 1776, Thomas Jefferson visited Sparhawk's stationery store in Philadelphia and purchased two items: for himself, a thermometer built by the London firm Nairne and Blunt, and for his wife, Martha, a pair of gloves. The next day he returned to buy a barometer.[1] Following a daily routine that he would maintain throughout his life, Jefferson then made detailed weather notations, entering that at 6:00 A.M. the temperature was 68 degrees; at 9:00 A.M., at the same time Congress was receiving the Declaration of Independence from the Committee of Five, 72.5 degrees; at 1:00 P.M., 76 degrees; and at 9:00 P.M., 73.5 degrees.[2]

Meteorological observations might not seem an obvious way to celebrate the advent of a nation, but for Jefferson, who was "the best-informed American of his day"[3] in meteorology, the accurate measurement of America's weather was integral to the scientific construction of this new nation. In Philadelphia, almost as an allegory, the Declaration of Independence was read from "the observatory platform behind the State House, which had been erected in 1769 to observe the transit of Venus."[4] Jefferson's famous declaration that "science is my passion, politics my duty"[5] may sound like a statement of alternatives, but for him and his contemporaries the two were mutually supportive. For the next fifty years he would lead the massive scientific project of defining America as a republican habitat, an environment whose climate and soil reflected its democratic institutions.

There were many reasons why accurate weather records were important to the new republic. Commerce and agriculture would benefit from a more comprehensive knowledge of its climate. Also with the resurgence of Hippocratic medicine in the 1660s, climate and weather had become fundamental to the medical understanding of the causes and geographical distribution of diseases.[6] Accurate weather records nevertheless served a very definite political purpose. Defining America as a geographical space required both the charting of its major physical features and boundaries (most notably, for Jefferson, the mapping of its Western interior) and the scientific description of its *climate*, a term that had become almost synonymous with "place." The American people were to be held together not only by the land that they settled but by the accurate scientific measurement of the air that they breathed and the presumed effects of heat and cold on their bodies, their productive and reproductive capacities, and their general health as a people.

In a January 1797 letter to Constantin Volney, Jefferson declares that if it had not been for the Revolutionary War he would have organized a major study of weather across the entire state of Virginia. "As I had then an extensive acquaintance over this State," he writes, "I meant to have engaged some person in every county of it, giving them each a thermometer, to observe that and the winds twice a day, for one year, to wit, at sunrise and at four P.M. (the coldest and the warmest point of the twenty-four hours), and to communicate their observations to me at the end of the year."[7] As early as 1778, he wrote to Giovanni Fabbroni that though much of his time "is employed in the councils of America," he wanted nevertheless to enter into a "philosophical" correspondence, particularly relating to climate:

> It might not be unacceptable to you to be informed for instance of *the true power of our climate* as discoverable from the Thermometer, from the force and direction of the winds, the quantity of rain, the plants which grow without shelter in the winter &c. On the other hand we should be much pleased with co[n]temporary observations on the same particulars in your country, which will give us a comparative view of the two climates.[8]

Climate study was implicitly a comparative activity because the merits of a region were understood in relation to others. Jefferson's goal was to demonstrate that America's climate could compete with anywhere else in the world in supporting the growth of healthy and strong people.

That Jefferson made science subservient to politics has not gone unnoticed by historians. John C. Greene, for instance, notes that

> to Americans living in this period of exploding scientific inquiry, the fundamental fact conditioning every thought and deed was the consciousness that they were now an independent nation. With respect to science this meant two things: as the example par excellence of useful knowledge, science must be cultivated to promote the interests, prosperity, and power of the rising American nation; and as the supreme example of the powers of the human mind, the successes of science challenged Americans to prove to the world that republican institutions were as favorable to intellectual achievement as they were to liberty.[9]

Meteorology played a part in the scientific construction of the republic, its people, and its technological future, just as it had served similar nationalistic purposes in other countries. In France, through the influence of the Abbé Du Bos, Montesquieu, the Abbé Raynal, Buffon, and Voltaire; in Germany, notably in the writings of Winckelmann, Herder, Lessing, and Georg Forster; and in England, in a wide range of political, literary, scientific, medical, and colonial writing, climate had become primary to the description and understanding of nations and peoples. The eighteenth and nineteenth centuries produced a politics of weather in which climate functioned as a primary element of cultural and political analysis. Arguments about the relative intellectual and political merits of nations were rarely voiced without some reference, implicit or otherwise, to the quality of a region's weather. Few writers, in short, would venture to discuss the history, politics, or cultural institutions of a country without first checking their meteorological instruments. The thermometer thus provided a scale for measuring the success of a people—"the true power of our climate."

The "politics of climate" was part of a broader environmental discourse whose role in the formation of European and American ideas about cultural and national differences has not been adequately recognized. At a time when monarchical governments and ecclesiastical institutions were in crisis, Europe witnessed the emergence of a highly politicized environmental discourse, one that defined the cultural and political identity of a people—their past, present, and future—in terms of their habitat. In antiquarian and artistic studies of the climate and soil of Ancient Greece; in scriptural studies of the ancient culture of the Hebrews and the topography of the Holy Lands; in political writings on the rugged health of the Swiss; in explorers' accounts of the perfect physical proportions of the people of Tahiti or the degenerate bodies of the people of Tierra del Fuego or Hudson's Bay; in colonial accounts of the dangers (both moral and physical) of the "tropics"; or in the wholesale celebration of the temperate qualities of Europe and its people; everywhere writers felt compelled to explain differences in societies and cultures in terms of differences in the world's physical environments. The conventional approach to the description of societies during this period—their division into hunting, pastoral, farming, and commercial societies—was not simply a description of a people's means of subsistence but also of the physical worlds that these societies respectively produced. As the narrator of Wordsworth's "Michael" declares, he loved shepherds of the Lake District, "not verily/For their own sakes, but for the fields and hills/Where was their occupation and abode."[10]

Environmental and political discourses were powerfully fused, as the description of physical environments was to a great extent a description of "political environments," that is to say, of places whose physical characteristics not only formed the character, lifestyles, and social and political institutions of the people who inhabited them but, in turn, were also produced by the actions and beliefs of a people. Climate, however, is as much an idea as it is a physical reality. In bringing the idea of an American climate into being, in producing the idea of a unified climate despite the climatic diversity of this vast region, and in arguing against ideologically competing versions of that reality being promulgated by European writers, American meteorology played an important role in producing the idea of the nation. Benedict Anderson's argument that nations are essentially "imagined communities" is relevant here because the idea that linked people together in this newly formed nation was that they shared the same political climate.[11]

An examination of the role of climate theory in the emergence of the concept of America as a political state provides access to certain aspects of the geopolitical dimensions of environmental discourse. It also helps clarify why writers during the Romantic period talked so much about the weather. For contemporary observers, America was a tremendous geographical experiment in the power of human beings to transform their physical environment and adapt it to their needs. Few discussions of the relationship between human beings and their environments or of the possibilities of human government proceeded without at least some passing mention of the enormous changes that were taking place there. Because America passed during this period from conceiving itself as a colony to conceiving itself as a nation, a history of the scientific construction of American weather provides an ideal opportunity to understand the manner in which politics and meteorology mutually supported the Romantic construction of national environments.[12]

Climate, Reproduction, and Colonial Biogeography

Jeffersonian meteorology developed as part of a strong nationalistic protest against the largely negative descriptions of America's climate that were then being promulgated in Europe. In France, notably in the writings of Buffon and the Abbé Raynal; in Germany, particularly through Cornelius de Pauw and later Hegel; and, in England, by Oliver Goldsmith and William Robertson, America was being portrayed as an inferior environment, as yet unfit for the healthy growth of plants or animals, and by extension human beings. For Hegel, the Americas constituted a "new world" not only in the recentness of their discovery, but also "in respect of their entire physical and psychical constitution."[13] For Hegel, "new" meant "immature," because it was land that had only recently emerged from the ocean floor and been populated by human beings. These ideas were articulated in the newly emerging field of biogeography (denominated "geographical history" by eighteenth-century naturalists), a field concerned with "the methodical study of facts relative to the distribution of animals and plants over the surface of the earth."[14]

Questions about the present and past distribution of plant and animal species across the globe had been raised before, especially in interpretations of the Biblical

Flood and Noah's Ark, but as a systematic field of inquiry, as the science of the factors influencing the restriction of communities of plants and animals to specific global habitats and of the mechanisms by which they were spread across the globe, biogeography was primarily indebted to the work of Linnaeus and Buffon. At a time when Europeans were laying the foundation for global empires, this new science was an important component of colonial science, as the study of the biological and environmental factors affecting the distribution of biota across the globe. The comparison of plants, animals, and habitats was never a neutral activity. The study of plant and animal migration, and of the long-term impact that a change of place had, or might have, on their physical characteristics and well-being, promised to provide clues to the viability of European expansionary migration. Much of the justification for expansion also came from this line of inquiry, in the frequently made argument that colonial regions were biologically weak habitats that needed to be improved through the introduction of European culture and technology.

Biogeographical discourse developed within the context of the enormous ecological transformations wrought by colonialism. This was the story not only of the migration of Europeans but also of the transportation and introduction of their foodstuffs and domestic animals to new regions of the world. Alfred A. Crosby notes that the immigrant did not come alone but "as part of a grunting, lowing, neighing, crowing, chirping, snarling, buzzing, self-replicating and world-altering avalanche."[15] The question of whether plants and animals could be adapted to new climates and soils was thus, by necessity, a very practical colonial concern from the fifteenth century onward. Even if early settlers paid scant attention to the possible long-term effects of climate change on their own bodies, they nevertheless had a very practical understanding of its impact on their food crops and livestock. European expansion was not an ecologically neutral event, but was built upon the efforts of settlers to reproduce the European ecosystem wherever they went. The American colonists were not just migrating to a new region of the world; they were seeking to re-create and perhaps even surpass the actual environment that they had left, to create a "second England" in the New World.

The emergence of biogeography in the eighteenth century constitutes an explicit acknowledgment that expansionary success could not be separated from the deliberate management and transformation of colonial environments to meet European needs. France and England were among the first to recognize the advantages that would accrue from establishing colonies that could produce foodstuffs and raw materials that could not be produced, or produced in sufficient quantity, in Europe. An extensive commerce in importing foreign plants for use in landscape gardening emerged during the early eighteenth century.[16] By its end, plant exchange had become an aspect of state policy, as was the case in William Bligh's ill-fated 1787 commission on the *Bounty* to introduce Tahitian breadfruit to the West Indies.[17] The eighteenth century made a business of ecological imperialism.

As the keeper of the Jardin du Roi, in Paris, Buffon occupied a position similar to Joseph Banks's at Kew Gardens. Both gardens functioned as commercial centers for the transportation of plants and animals to new regions of the globe for the benefit of England and France. David Mackay argues, for instance, that from the sixteenth century onward,

> apart from the Spanish imports of bullion from their South American dominions, the wealth of colonies lay in their tropical vegetable products—spices, sugar, tobacco, dyes and medicinal drugs, tea and coffee, oils. Some of these new vegetable products could be introduced into England; vegetables such as potatoes, tomatoes and spinach to supplement the diet; exotic trees such as the rhododendron, plane and cedar to grace the parklands of the nobility and gentry. In such cases, botanical gardens provided reception centres where plants could be acclimatised, propagated and studied. Some plants, of course, would not grow in temperate climates, but many could be transferred from one tropical country to another. All European countries with Asian or American empires attempted the interchange of tropical plants to their dominions, the French being particularly active.[18]

Both France and England were constructing "colonial ecologies," regions that either duplicated the European environment and would support extensive immigration or tropical regions whose environments could be administered to supply products that could not be naturalized in a temperate climate. Knowledge of the distribution, migration, and adaptation of plant and animal species and of the relationship between biological form, function, and place was thus of practical commercial importance to nations engaged in exploiting environments on a global scale and in expanding their influence beyond traditional national boundaries.

Even as the primary scientific focus of biogeography was on the natural or artificial distribution of plants and animals and their adaptation to new environments, its most controversial aspect lay in its reflection on the biological relationship between people and environments. One of the more radical gestures of eighteenth-century naturalists, from Linnaeus onward, was to argue that human beings, though impressive as a species, nevertheless were part of natural history. Their diverse characteristics could thus be seen as a geographical phenomenon. Faced with the uneven distribution of human beings across the globe, it was reasonable to ask whether differences between people reflected differences in the environments that they inhabited, with the corollary question of whether human beings were actually biologically capable of migrating or adapting to different geographic habitats. If plants and animals were restricted to specific environments, it was reasonable to assume that perhaps geography and environment might play just as significant a role in the distribution of human beings across the globe. Biogeography thus broached questions of far-reaching importance to European colonists, especially as they attempted to settle in regions that were substantially different from the temperate zones of the globe in which they had previously been so successful.

Since Buffon provided the general theoretical framework of this debate during the latter part of the eighteenth century, his work is important for understanding the interaction of biogeography, politics, and climate.[19] Like other eighteenth-century naturalists, he followed Linnaeus in regarding climate as the primary biological boundary mechanism by which plant and animal species were restricted to specific habitats. Climate was the means by which each organism found its preestablished place within a carefully ordered universe. Whereas Linnaeus understood the influence of climate on the geographical distribution of plant and animal biota in general terms, Buffon refined this model by arguing that it exerted its influence specifically in regard to the sexual reproduction of species. For Buffon, the primary unit of biological life was not the individual but the species, which was defined by reproduction. "Every species of animal is distinguishable from another by copulation," he writes. "Those may be regarded as of the same species which, by copulation, uniformly produce and perpetuate beings every way similar to their parents; and those which, by the same means, either produce nothing, or dissimilar beings, may be considered as of different species."[20] Two animals belong to the same species, then, not because they look alike, but because they have the ability to produce progeny like themselves that are themselves capable of passing on that likeness through reproduction. It is "this faculty of producing beings similar to themselves, this successive chain of individuals which constitutes the real existence of the species" (*NH* 2:16). One consequence of this reconceptualization of life in terms of reproduction was that Buffon shifted biogeographical concern away from the effects of soil and climate on the individual plant or animal toward their long-term impact on the ability of a species to reproduce its likeness over time. Climate debate, therefore, increasingly centered on its impact on sexual reproduction, on the life of a population of like beings—a species—rather than on the life of the individual, which was now primarily a medical concern.

For Buffon, species do not actually change because every existing plant or animal is only a copy—a reproduction, in the strongest sense—of the original form of the species. A change in climate, therefore, cannot change the species form. What it can radically alter, however, is the ability of a species to reproduce that form adequately. Buffon used the term *degeneration* to describe this failure to reproduce the species. Not surprisingly, colonial history provided him with numerous instances of this process, notably in the many instances in which European settlers had been unsuccessful in naturalizing domesticated plants and animals to new geographical regions. A new environment, Buffon argued, has a relatively minor impact on the animals that are first transported to it because their mature bodily form would have already been established. However, weakened by a new climate and by a change in nutrition, these animals would pass on these weaknesses to their progeny which would themselves labor under the increased impact of climate as they developed. This generation would in turn pass on their parents' weaknesses, now increased by their own, to the next generation. Sexual reproduction thus magnifies the effects of environment, so much so that eventually, over the course of many generations, the original form of a species can eventually become almost unrecognizable, not because it has been lost but because the species has lost the physical ability to reproduce or copy it. For Buffon, species do not disappear from the earth suddenly, through cataclysms, but slowly, as environmental factors attack their ability to reproduce sexually the original mold that defines a population. Buffon's biogeography thus draws a powerful link between sexuality and the environment, suggesting that colonial regions do not immediately register themselves on the bodies of newly introduced species (including human beings), but instead do so over time, in producing a progressive decline of their power to reproduce themselves. Through Buffon, the relationship between colonialism and climate was given an explicit sexual turn. It boiled down to a question of the long-term continuance of animals in a new environment that progressively destroyed their sexual ability to reproduce themselves.

The centerpiece in Buffon's biogeography was provided by the entire Western hemisphere—by "America." When comparing the animals of the New World with those of the Old, he made what he considered to be an extraordinary discovery: the quadrupeds of the New World seemed to be invariably smaller and weaker than those of the Old World. North America had no elephants (at best the piglike tapir), no giraffes, no hippopotamuses, and no rhinoceroses; no tigers, panthers, or lions, only the puma, which was smaller, weaker, and more cowardly than the lion; and the best camel that the American environment could come up with was the South American llama (called by some "*sheep*, and by others *camels* of Peru") (*NH* 5:213). Not only were they smaller, but also the number of species of quadruped was fewer, suggesting that the New World lacked not only reproductive power but also creativity.

Buffon's "America" was Darwin's Galapagos Islands writ large: a region where the very paucity of biological life-forms provided an insight into the fundamental relationship between climate and species forms. Assuming that "all the animals of the New World were originally the same as those of the Old," Buffon argued that the animals of the New World had degenerated through "all the effects of a climate that had become new to them" (5:271). This negative picture of America as a biological environment also applied to the domestic animals that the Europeans had brought with them to the New World:

> The horses, donkeys, oxen, sheep, goats, pigs, dogs, all these animals, I say, became smaller there; and . . . those which were not transported there, and which went there of their own accord, those, in short, common to both worlds, such as wolves, foxes, deer, roebuck, and moose, are likewise considerably smaller in America than in Europe, and *that without exception*. (5:250, my emphasis)

Even when Buffon was willing to admit that some domestic animals, such as sheep, goats, and hogs, seemed to be even more prolific in America than in Europe, he denied that, with the exception of the pig, they were of the same quality. The ewes, he claimed, "are commonly more meagre and their flesh less succulent and tender than that of the European sheep," and "the flesh of the ox . . . is not so good at Brasil as in Europe" (5:221, 223). There seemed something in America's environment that was hostile to the strong, healthy development of animal life. For Buffon, who followed Hippocratic medicine in believing that heat was the essential factor in successful sexual generation, whereas cold and humidity were detrimental, the small size and comparative weakness of New World animals was a zoological confirmation of the biological inadequacy of the American climate.[21] In fact, the only animals that seemed to prosper in this wet, swampy environment were cold-blooded, the many reptiles and insects, which grew there to be enormous. "The largest spiders, beetles, caterpillars, and butterflies, are found in Cayenne and other neighbouring provinces," Buffon writes. "The toads, the frogs, and the other animals of this kind, are likewise very large in America." In short, America's power as a habitat was "limited to the production of moist plants, reptiles, and insects, and can afford nourishment only to cold men and feeble animals" (5:253, 257).

Buffon's was hardly an objective biogeographical description of America. Instead, it established the European ecosystem as the standard by which other environments were to be measured. From such a perspective, America was an inferior environment where "animated Nature is weaker, less active, and more circumscribed in the variety of her productions" (5:237). Such a view ran counter to American colonial experience and to an older pastoral myth that had sustained America from its beginnings. Annette Kolodny indicates the degree to which European settlers in America acted out upon the landscape a psychosexual dream of "making it with paradise": "America's oldest and most cherished fantasy" is that of "a daily reality of harmony between man and nature based on an experience of the land as essentially feminine—that is, not simply the land as mother, but the land as woman, the total female principle of gratification—enclosing the individual in an environment of receptivity, repose, and painless and integral satisfaction."[22] European biogeography, as it was practiced by Buffon and his followers, challenged this popular gendered image of the American environment as a warm and fertile mother with an alternative, equally psychosexual representation of America as a "frigid woman." Alexander von Humboldt, in his caustic response to such simplifications, rightly caught the gendered aspect of this scientific myth: "America is a female figure, long, slender, watery, and freezing at 48°."[23]

Yet nature for Buffon was not wedded to its primordial frigidity; it could be improved through human intervention. Because climate was seen as the primary obstacle standing in the way of European expansion, it should not be surprising that one of the central strands of eighteenth-century biogeographical debate was the degree to which human beings are capable of struggling against the limits imposed on them by climate. It was a commonplace that human beings, unlike most other animals, had the capacity to adapt themselves to a wide range of physical environments. Through culture and technology, they had found the means of mastering all climates and transforming them, by clearing forests, draining swamps, and planting crops. From this perspective, the inferiority of the American environment and its plants and animals could be seen as a symptom of the failure of its indigenous people to actively modify it to meet their needs. In a context in which landscapes mirror the capacities of the people who inhabit them, the harshness of America's weather and the weakness, decrepitude, and sexual impotence of its biota were symptoms of a deficiency in its indigenous people. "Man makes no exception to what has been advanced," Buffon writes:

> In this New World, therefore, there is some combination of elements and other physical causes, something that opposes the amplification of animated Nature: there are obstacles to the development, and perhaps to the formation of large germs. Even those which, from the kindly influences of another climate, have acquired their complete form and expansion, shrink and diminish under a niggardly sky and an unprolific land, thinly peopled with wandering savages, who, instead of using this territory as a master, had no property or empire; and having subjugated neither the animals nor the elements, nor conquered the seas, nor directed the motions of the rivers, nor cultivated the earth, held only the first rank among animated beings, and existed as creatures of no consideration in Nature, a kind of weak automatons, incapable of improving or seconding her intentions. She treated them rather like a step-mother than a parent, by denying them the invigorating sentiment of love, and the strong desire of multiplying their species. For, though the American savage be nearly the same stature with men in polished societies, yet this is not a sufficient exception to the general contraction of animated Nature throughout the whole continent. In the savage, the organs of generation are small and feeble. He has no hair, no beard, no ardour for the female. . . . It is easy to discover the cause of the scattered life of savages, and of their estrangement from society. They have been refused the most precious spark of Nature's fire. They have no ardour for women, and, of course, no love of mankind. Unacquainted with the most lively and most tender of all attachments, their other sensations of this nature are cold and languid. . . . Hence no union, no republic, no social state, can take place among them. . . . Their heart is frozen, their society cold, and their empire cruel. (*NH* 5:251–52)

The American native, in Buffon's estimation, has suffered the same biological degeneration as other warm-blooded animals in the New World, and this degeneration is expressed in sexual physiology, in generative organs that are "small and feeble," in the absence of "beard[s]," and in a lack of "ardour for the female." It seems that it is this very deficiency of sexual desire, the American native's inability to master either the coldness of his own body or his exterior environment, that has prevented him from progressing beyond the savage state. "This indifference to the sex," Buffon argues, "is the original stain which disgraces Nature, prevents her from expanding, and, by destroying the germs of life, cuts the root of society." In this image of castration, of society cut at its root, Buffon establishes a fundamental link between sexual desire, population growth, technological progress, and political institutions. "Nature, by denying him the faculty of love, has abused and contracted him more than any other animal" (5:252–53). For Buffon, population is power, and it is the failure of the indigenous peoples to populate America's wilderness that has prevented them from dominating it. Frozen in time, because they are frozen in body, these people are too weak to change their environment and they are denied the power that comes from political institutions. "No union, no republic, no social state," Buffon argues, "can take place among them."

Buffon's representation of the relationship between the indigenous peoples and the physical landscape is also typical of late Enlightenment environmental thought in its structuring of ecological transformation in terms of male and female reproduction. Equating men with technology and power, and women with nature, Buffon argues that the essential problem with America prior to its being settled by Europeans is that it has long been locked in an infertile, frigid relationship, a state indicated by its raw natural state and lack of population. In America, he writes:

> Nature remains concealed under her old garments, and never exhibits herself in fresh attire; being neither cherished nor cultivated by man, she never opens her fruitful and beneficent womb. Here the earth never saw her surface adorned with those rich crops, which demonstrate her fecundity, and constitute the opulence of polished nations. In this abandoned condition, every thing languishes, corrupts, and proves abortive. (5:257)

America is described as a land that is still unfamiliar with the pleasurable warmth of man's civilizing touch, and consequently, her frozen loins can only yield people deficient in the human drive to populate and dominate the earth. The homology drawn between nature as female and the native's assumed lack of "ardour for the female" is quite obvious, for Buffon draws on the traditional idea that conception cannot take place without female orgasm, without the heat and pleasure that encourage the womb to open in order to grasp the male seed.[24] Drawing on the popular eighteenth-century association of sex with agriculture, he represents America as an unadorned and "unbroken" "virgin" land, while the European landscape is presented in a "dress" that demonstrates her fertility, itself an expression and product of the power derived from population. In contrast to Thomas Malthus, the power of population does not produce scarcity, but instead wealth. "The scarcity of men . . . in America," he writes, "is the chief cause why the earth remains in a frigid state, and is incapable of producing the active principles of Nature" (*NH* 5:257).

By claiming that the "humanness" of human being is expressed in the mastery of physical environments, in enfranchisement from the boundary mechanisms of climate and soil, Buffon thus provides a biological and demographic model for differentiating and evaluating human cultures. It is a scheme that devalues native societies, because the closer a people are to nature, the less they are able to escape the restraints placed on their bodies and minds by the physical environment—and nature strives to keep animals in their place. As Goldsmith argued in *The History of Earth and Animated Nature* (1774), a book largely derived from Buffon, "it is no unpleasing contemplation to consider the influence which soil and climate have upon the disposition of the inhabitants, the animals and vegetables of different countries. That among the brute creation is much more visible than in man, and that in vegetables more than either."[25]

For Buffon, the indigenous population of America had shown little more success in modifying the effects of climate on their bodies than had its native biota. To be controlled by climate is to be closer to animals and plants than to civilized human beings. The laws developed by the biogeographer can be said, then, to apply more fully to primitive societies than to advanced ones, where the effects of climate and soil can be ameliorated through trade and commerce or through the creation of alternate climates, the greatest of these being, as Byron suggests, "in-door life," "that fair clime which don't depend on climate."[26] At the same time, Buffon values those actions that allow human beings to control and "improve" their environments. Civilization, as the fullest expression of *human* being, is thus identified with the forces that free human beings from the control of the places they inhabit. Geographical mobility, as an expression of the power to control and remake the boundary mechanisms of climate and soil, is thus valued as an index of civilized life. This argument, revised and adapted as needed, can thus be said to have provided the intellectual foundations of empire, of a colonial activity that saw itself not only as adapting itself to different global environments but also as providing the technological and cultural means for transforming them. In Buffon's biogeography, colonial ecologies display the marks of their "dehumanization," and they must be remade to meet the needs of human beings who define themselves in their capacity to transform global environments. What are the ultimate ends and justification of European global expansion? a world in which the extremes of colonial weather have been fully moderated to meet human needs.

In ways that have not been adequately recognized, Buffon's biogeography places eros at the heart of the transformation of colonial landscapes. "Love" is the power that populates and thus transforms colonial natures into their civilized alternatives. This aspect of Buffon's biogeography is quite clear in the *Natural*

History of Birds in which he draws attention to the "singular" fact "that in all populous and civilized countries, most of the birds chant delightful airs, while, in the extensive deserts of Africa and America, inhabited by roving savages, the winged tribes utter only harsh and discordant cries, and but a few species have any claim to melody."[27] There are no nightingales in the mournfully silent continent of America. The feathers of its birds may be beautiful, as if nature sought to exhaust "all the rich hues of the universe on the plumage of the birds of America, of Africa, and of India" (1:13), but they are notably inferior in song. With the exception of the mockingbird, they can only utter "hoarse, grating, or even terrible cries." Here Buffon turns America's birds into ornithological savages, whose bright feathered dress comes at the expense of the development of harmonious language and song. The birds of temperate regions are less colorful, "stained with lighter and softer shades," but they can claim the higher privilege of song. Modern zoology would insist that there is no connection between civilization, population, and the development of birdsong, but for Buffon, like many of his contemporaries, the boundary between nature and culture is not fixed, only an interactive interface that allows human beings to play a key role in the improvement of nature. The "real alteration which the human powers have produced on nature," he writes, "exceeds our fondest imagination: the whole face of the globe is changed; the milder animals are tamed and subdued, and the more ferocious are repressed and extirpated. They imitate our manners; they adopt our sentiments; and, under our tuition their faculties expand" (14). For Buffon, birdsong is "partly natural, partly acquired," so he concludes that American birds display only rudimentary powers of song because they lack adequate models of *human* language to imitate. The absence of birdsong in the New World is thus, at least on one level, another indictment of native peoples, who have given birds not only a world that offers them little to sing about but also inadequate linguistic models for song.[28]

In keeping with the primacy that Buffon accords sexuality in the development of culture, he concludes, not surprisingly, that the absence of sophisticated song is also owing to a lack of sexual virility in America's birds. Buffon had little doubt that "love" incited the birds to song, for they "cease entirely" or lose "their sweetness" after the breeding season. He went further, however, to speculate that there was a possible "physical relation between the organs of generation and those of voice." Noting that in humans puberty brings changes in both the reproductive organs and the larynx, Buffon suggested that birds during the breeding season might undergo similar changes: "the testicles, which in man and most of the quadrupeds remain nearly the same at all times, contract and waste almost entirely away in birds after the breeding season is over, and on its return they expand to a size that even appears disproportioned."[29] Could a similar physical enlargement be taking place in birds' vocal organs, he queried? Buffon would also suggest that the harsh, discordant, low songs of American birds were like those of birds after the mating season is over. Despite their glorious plumage, perhaps they had not yet been awakened to the powerful incitements of physical desire. The immaturity of America as a continent was thus extended to its creatures. Any advance in their procreative capacity was owing to the advance of civilization. The warming touch of European expansion extended not only to the land but its creatures.

Many of Buffon's contemporaries adopted his portrayal of America as a physical habitat whose biological inadequacies were expressive of its status as a colonial environment. De Pauw presented the most devastating portrait of America, arguing that "it is without doubt a grand and terrible spectacle to see a part of this globe so disgraced by nature that everything there is either degenerate or monstrous." De Pauw went beyond biology to claim that America's lack of virility even extended to its metals: its iron, he claimed, is "infinitely inferior to the iron of our continent, so that one could not even make nails out of it."[30] Intoning in the *Philosophy of History* that "the true theatre of History is . . . the temperate zone," Hegel would also claim that "America has always shown itself physically and psychically powerless." The "northern birds are drabber, but better songsters . . . witness the nightingale and the lark, which are not to be found in the tropics," he writes, and goes on to cite the prophecy of a contemporary travel writer, "that once the almost inarticulate grunts of degenerate men have ceased to sound through the forests of Brazil, many of the feathered songsters there will pour forth finer melodies."[31]

In England, Buffon's ideas were popularized by Lord Kames in *Sketches of the History of Mankind* (1776), William Robertson in *The History of America* (1777), and Oliver Goldsmith in the *History of the Earth and Animated Nature*.[32] Though Wordsworth and Coleridge did not adopt Buffon's negative view of America, partly because of the powerful effect that reading John Bartram's *Travels* (1791) had upon their understanding of the region, many of the British Romantic poets were profoundly influenced by Buffon's biogeographical ideas. In William Blake's *America* (1793), revolution is portrayed not only as a violent rape of a "nameless female" which produces language, but also in psychosexual terms as a "fierce embrace" that brings fertility to a frigid land:

> On my American plains I feel the struggling afflictions
> Endur'd by roots that writhe their arms into the nether deep:
>
> . . . .
> Oh what limb rending pains I feel. Thy fire & my frost
> Mingle in howling pains, in furrows by they lightnings rent.
> This is eternal death; and this the torment long foretold." (Pl. 2:10–17)[33]

As I have elsewhere noted, Percy Bysshe Shelley's geographical ideas were deeply informed by Buffon, whom he first read in 1811.[34] In *Queen Mab* (1813), Shelley presents a devastatingly negative picture of Northern natives as a degenerate population, who, living in the "gloom of the long polar night" and amid the "snow-clad rocks and frozen soil,"

> Shrank with the plants, and darkened with the night,
> His chilled and narrow energies, his heart,
> Insensible to courage, truth, or love,
> His stunted stature and imbecile frame,
> Marked him for some abortion of the earth,
> Fit compeer of the bears that roamed around,
> Whose habits and enjoyments were his own:
> His life a feverish dream of stagnant woe.[35]

Even more significant was Buffon's impact upon the structure of *Prometheus Unbound* (1820), where political revolution is registered in a change of climate. Prometheus's curse, though forgotten by him, is preserved materially by a nature whose once temperate climate is now riven by "alternate frost and fire" (1.268). With the revocation of the curse, Prometheus ushers in the possibility of an ecological revolution that is announced by the recovery of seasonal change. Shelley grasped the essence of Buffon's biogeography, even as he undercuts its overt imperial resonances, for he makes the love of Prometheus and Asia the force that makes this transformation possible. Mary Shelley also knew Buffon's *Natural History*, which she read in June and July 1817 while writing *Frankenstein*.[36] Buffon and Pliny are favorites of Victor Frankenstein, for he continues to read them long after he has discarded the work of the alchemists Cornelius Agrippa, Albertus Magnus, and Paracelsus.

Mary Shelley's monster also enters into the debates opened up by Buffon. In his account of his early development he declares that among his earliest sensory discoveries was "that a pleasant sound, which often saluted my ears, proceeded from the throats of the little winged animals who had often intercepted the light from my eyes." His natural response is to seek "to imitate the pleasant songs of birds, but . . . the uncouth and inarticulate sounds which broke from me frightened me into silence again." With more knowledge he learns "that the sparrow uttered none but harsh notes, whilst those of the blackbird and thrush were sweet and enticing."[37] At this point in his history, the monster's ideas of beauty and song are limited to the models provided by nature, so it is with a sense of wonder that he hears music for the first time: "The young girl was occupied in arranging the cottage; but presently she took something out of a drawer, which employed her hands, and she sat down beside the old man, who, taking up an instrument, began to play, and to produce sounds, sweeter than the voice of the thrush or the nightingale." The monster is enthralled by the beauty of Agatha and the sweetness of her father's music, but this experience also leaves him feeling even more deeply his status as an outsider—a "poor wretch." Upon hearing Felix read, he learns that human beings also make "sounds that were monotonous" compared to "the old man's instrument or the songs of the birds," and he harbors the hope that by imitating these sounds, by learning language, he can become more like these people and perhaps enter their community.[38] On failing to do so, recognizing the boundaries that separate him from European culture, he concludes that he is better suited for life in South America.

Buffon's biogeography gave new meaning to the traditional metaphoric association of poetry and avian song because it would seem that the obstacles that lay in the way of the development of birdsong in the New World also applied to poetry. In this regard, though Goldsmith's "The Deserted Village" (1770) is primarily concerned with deploring the depopulation of British rural life and the poetry that it supported, it played an important role in questioning the poetic merits of the American habitat. Goldsmith describes America as a "horrid" place, where "silent bats in drowsy clusters cling," where "poisonous fields" produce only "rank luxuriance," and where the stranger can expect to encounter only scorpions, rattlesnakes, tigers, and "savage men." Instead of the "breezy covert of the warbling grove," there are only "matted woods where birds forget to sing."[39]

Through his reading in Buffon, Robertson, and Goldsmith, John Keats was already biased toward America, but this attitude was exacerbated in 1818 by his brother George's emigration to Kentucky and the financial difficulties that followed. Writing in October of that year to his sister-in-law Georgiana, Keats would assert that "those Americans are great but they are not sublime Man—the humanity of the United States can never reach the sublime." In the same letter, he nevertheless entertains the notion that Georgiana, pregnant at the time, might soon give birth to "the first American poet," an idea that gives rise to the lovely lullaby "'Tis the 'witching time of night'" in which Keats prophesies his appearance: "Bard art thou completely!/Little child/O' the western wild,/Bard art thou completely!"[40] In "What can I do to drive away," a much darker image of the American landscape emerges, probably influenced by Keats's anxious awareness of his isolation in the face of the onset of consumption:

Where shall I learn to get my peace again?
To banish thoughts of that most hateful land,
Dungeoner of my friends, that wicked strand
Where they were wreck'd and live a wretched life;
That monstrous region, whose dull rivers pour
Ever from their sordid urns unto the shore,
Unown'd of any weedy-haired gods;
Whose winds, all zephyrless, hold scourging rods,
Iced in the great lakes, to afflict mankind;
Whose rank-grown forests, frosted, black, and blind,

Would fright a Dryad; whose harsh herbaged meads
Make lean and lank the starv'd ox while he feeds;
There flowers have no scent, birds no sweet song,
And great unerring Nature once seems wrong.[41]

Unsatisfied with depicting America as a "monstrous" region whose "afflict[ing]" climate and "harsh" vegetation starve the domestic animal even "while he feeds," and where "birds [have] no sweet song," he adds a particularly Keatsean item to what was by now a lengthy catalog of complaints against this habitat—in America even the "flowers have no scent." On January 15, 1820, Keats had another opportunity to comment on the New World. George, who had recently returned to England to settle financial matters, was occupied in "making copies of my verses—He is making now one of an Ode to the nightingale." Though sunny, the day was extremely cold, and Keats must have felt the contrast between this January morning and the idea of a "warm South"—the world of "Flora," "country green," and "sunburnt mirth"—that had served as the inspiration, "the blushful Hippocrene," of the "Ode to a Nightingale." He wrote to Georgiana that on such a day, "an American one," the experience of reading a poem such as this one could only be compared to "reading an account of the b[l]ack hole at Calcutta on an ice bergh."[42]

Buffon's attack on America's climate seemed to support an anticolonial agenda. Horace Walpole drew this conclusion when, in a letter, he commented: "Buffon says, that European animals degenerate across the Atlantic; perhaps its migrating inhabitants may be in the same predicament."[43] Gilbert Chinard has noted that de Pauw's *Recherches* served a similar purpose in discouraging German emigration to America,[44] while William Robertson's *History of America* was also motivated by a similar concern. Certainly, it was this aspect of Buffon's biogeography that was most disturbing to the leaders of the new republic, then engaged in the major task of encouraging immigration. Yet it is important to recognize that Buffon was not just describing the environment of America: he was providing a biogeographical theory that would justify colonial efforts to change it. It was not climate that stood in the way of its successful colonization, but the absence of a sufficient population to transform this New World. "Nothing appears to be more difficult, not to say impossible, than to oppose the successive cooling of the earth, and to warm the temperature of a climate," writes Buffon, "yet this feat man can and has performed" (*NH* 5:336). Underlying his negative representation of the climate of America as a physical boundary preventing the importation of plant and animal species was an extraordinary optimism about Europeans' ability to change it to meet their needs. For Buffon, the inferiority of America's climate was a sign of its colonial status, a justification for European expansion into America, to shape the landscape into what Nature was intended to be, that is, into another Europe. Buffon was among the first to recognize that America was undergoing what Carolyn Merchant has called an "ecological revolution."[45] The American settlers, in little more than a century, had accomplished changes that had taken Europeans one thousand years. "Some centuries hence," Buffon writes, "when the lands are cultivated, the forests cut down, the courses of the rivers properly directed, and the marshes drained, this same country will become the most fertile, the most wholesome, and the richest in the whole world, as it is already in all the parts which have experienced the industry and skill of man" (5:260).

As one of the dominant expressions of European colonial biogeography, Buffon's work can be read as a powerful expression of the ideological link that was then being formed between nation building and the transformation of physical environments. Buffon is very much a product of Enlightenment in his belief that through government, population, and technology human beings do not simply dominate or control nature but actually assist it to become what it ought to be: an environment perfectly suited to human needs. Mankind "wonderfully seconds" nature in her "operations, that by the aid of his hands, her whole extent is unfolded, and she has gradually arrived at that point of perfection and magnificence in which we now behold her" (5:333). It is a gendered theory of culture, modeled on husbandry, in both a sexual and cultural sense, for it likens the cultivation of the earth to sexual reproduction. Ann Bermingham, in her study of English rustic landscape painting, has suggested that one reason why these paintings achieved such popularity between 1750 and 1850 is that they succeeded in "naturalizing" English class society, making the landscape garden, a cultural product of the large landowners of the eighteenth century, into nature: "during the period of accelerated enclosure, the garden and 'common nature' were no longer seen in opposition but in symbiosis."[46] Buffon's biogeography provides a far grander naturalization of the culture of the European landed gentry because it suggests that only "civilized" environments can truly fulfill Nature's desire to be fertile, warm, beautiful, and richly dressed. And just as the "ideology of landscape" justified the enclosure of land in England, so too, Buffon's biogeography justified colonial expansion—the conception of a "global garden." The Old World was made the template for all other geographic environments and the people who inhabited them. "Compare rude with cultivated Nature," Buffon writes.

> Compare the small savage nations of America with those of our civilized people, or even with those of Africa, who are only half cultivated. Contemplate the condition of the lands which those nations inhabit, and you will easily perceive the insignificance of men who have made little impression on their native soil. Whether from stupidity or indolence, these brutish men, these unpolished nations, great or small, give no support to the earth; they starve without fertilizing her; they devour every thing, and propagate nothing. (*NH* 5:333–34)

Nature not only needs but wants to be civilized. *Wild nature, barren nature, unimproved nature, starving nature, frigid nature*—these are the terms that Buffon uses to describe colonial environments. For him, European colonization was analogous to a creative breeze that warmed a female earth and made her capable of receiving seed.

From Colony to Nation: Climatic Change in Republican America

American natural history was formulated within the context of this sophisticated discourse on climate, which defined nations and peoples by the respective qualities of their environments and by the degree to which they were able to transform them to meet their needs. Jefferson led "the American contingent" in what was to become "an international rivalry over climate" as he sought to counter, in his words, a "theory in general very degrading to America" and to demonstrate scientifically that Buffon was fundamentally misinformed about the climate of the United States.[47] In a 1786 letter to Benjamin Vaughan, he discussed the manufacture of a hygrometer that would allow for a "comparison between the humidities of different climates," because the "opinion pretty generally received among philosophers that the atmosphere of America is more humid than that of Europe" was one of the pillars on which Buffon "builds his system of the degeneracy of animals in America" (*Papers* 10:646). He also sought to turn the tables on England, claiming to Constantin Volney in 1805 that America's "cheerful" climate "has eradicated from our constitutions all disposition to hang ourselves, which we might otherwise have inherited from our English ancestors."[48] Jefferson also placed his home state, Virginia, at its center—"the Eden of the U. S. for soil, climate, navigation & health." "There is not a healthier or more genial climate," he claimed, "in the world."[49]

In addition to meteorological records, Jefferson had recourse to America's animal life in his effort to demonstrate the superiority of its climate. Shortly after arriving in Paris in 1785, he showed Buffon the skin of a puma hoping (unsuccessfully) to correct the negative estimate of its size. In 1787, at great expense, he shipped from New Hampshire to Paris the skin, bones, and antlers of a North American moose, along with samples of the horns of North American elk, caribou, and deer. Because of numerous delays, the moose was in terrible shape even before it left New Hampshire, so when it finally did arrive in Paris, having lost most of its hair, it was undoubtedly a gift that Buffon might rather have declined. For Jefferson, this moose was a key piece of biological counterpropaganda for the new republic. Through it, he hoped to lay to rest any notion that North American quadrupeds could not compete with those of the Old World. At a time when America depended on immigration to bolster its population, he was not going to allow what he called "the lies of Paw, the dreams of Buffon and Raynal, and the well-rounded periods of their echo Robertson" to affect its future (*Papers* 13:397).

Jefferson's *Notes on the State of Virginia* (1785), which is one of the first major American regional natural histories, exemplifies the manner in which natural history in its capacity to conceptualize physical habitats was part of larger geopolitical debates about the relative merits of nations. After an initial discussion of the primary physical topography and minerals of the region, Jefferson turns to the biota of Virginia, providing an extensive, detailed tabular comparison of the respective weights and numbers of quadrupeds in both the Old and New Worlds, concluding that the New World compares favorably with the Old as a natural habitat. Even so, Jefferson was holding in reserve a biological ace in the hole: though the Old World might boast of its elephants, America could claim an even larger land beast—the hairy mammoth. Although only fossilized remains of this animal had been found in either the Old or the New Worlds, Jefferson felt that somewhere in the vast unexplored regions of the American Northwest mammoths still roamed the earth.[50] Such an animal, he claimed, "should have sufficed to have rescued the earth it inhabited, and the atmosphere it breathed, from the imputation of impotence in the conception and nourishment of animal life on a large scale." Anticipating his reader's rejoinder, "why [do] I insert the Mammoth, as if it still existed?", Jefferson asks, "in return, why I should omit it, as if it did not exist? Such is the oeconomy of nature that no instance can be produced of her having permitted any one race of her animals to become extinct; of her having formed any link in her great work so weak as to be broken."[51]

In 1797, as president of the American Philosophical Society, Jefferson presented a paper describing the discovery of the fossilized remains of what seemed to be an enormous American carnivore, the "megalonyx" or "megathere." (The animal was, in fact, a large ground sloth.) Jefferson also questioned Buffon's representation of America's native people. It is "an affecting picture indeed," he writes, "which, for the honor of human nature, I am glad to believe has no original. . . .The Indian of North America . . . is neither more defective in ardor, nor more impotent with his female, than the white reduced to the same diet and exercise" (183–84). Even as he admits that environmental factors, such as "diet and exercise," play an important role in differentiating native Americans from their European counterparts, he nevertheless claims that there are no fundamental biological differences: they "are formed in mind as well as in body, on the same module with the 'Homo sapiens Europaeus'" (187). That the political debate over the merits of American bodies extended to colonists of America is clear from a anecdote that William Carmichael recounted of a dinner party that he had attended in Paris. When Benjamin Franklin was asked his opinion of the Abbé Raynal's belief that "the human race had degenerated by being transplanted to another section of the Globe," he replied that to reach a satisfactory answer one had only to look round the table. "There was not one American present who could not have tost out of the Windows any one or perhaps two of the rest of the Company" (*Papers* 12:241).

Despite Jefferson's explicit criticism of Buffon's biogeographical representation of America as a habitat, the differences between the two men were not as great as they might seem; instead of rejecting the French naturalist's assertion of the link between the quality of a region's environment and the quality of its government, Jefferson agreed that America's climate was not a simple given but also reflected the progressive power of its republican institutions. His goal was not simply to document America's weather, but instead to establish scientifically that European settlement had produced a major change in the climate of America. "A change in our climate . . . is taking place very sensibly," he writes in *Notes on the State of Virginia*: "Both heats and colds are become much more moderate within the memory even of the middle-aged" (206–07). Indeed, climate change was one of the most talked about subjects during the early life of the republic. As Volney observed, in 1804,

> for some years it has been a general remark in the United States, that very perceptible partial changes in the climate took place, which displayed themselves in proportion as the land was cleared. . . . On the Ohio, at Gallipolis, at Washington in Kentucky, in Francfort, at Lexington, at Cincinnati, at Louisville, at Niagara, at Albany, everywhere the same changes have been mentioned and insisted on: longer summers, later autumns, shorter winters, lighter and less lasting snows and colds less violent were talked of by everybody; and these changes have always been described in the new settled districts, not as gradual and slow, but as quick and sudden, in proportion to the extent of cultivation. . . . [There can be] no room to doubt the truth of a sensible change in the climate.[52]

Jefferson had little doubt that the speed with which the new republic's climate had changed directly expressed the vitality of its people and their political, social, and scientific institutions. In a letter written to Lewis C. Beck in 1824, he describes the *Notes on the State of Virginia* as "the first attempt, not to form a theory, but to bring together the few facts then known" about climate, and declares that proper studies needed to be done "to show the effect of clearing and culture towards changes in climates."[53]

For Jefferson, American meteorology had a great story to tell, an epic narrative of the historical progress of its climate, a history of change profoundly linked to the history of its peoples. Throughout most of his life, as John Greene has noted, "Jefferson was his own weather bureau before such facilities existed, keeping daily records of temperature, wind, and rainfall."[54] Nevertheless, as he noted in 1822: "Of all the departments of science, no one seems to have been less advanced for the last hundred years than that of meteorology."[55] By then, however, through Jefferson's influence, an institutional structure had been set in place that would provide the scientific documentation for this story. He witnessed the beginning, between 1814 and 1818, of the National Weather Bureau by the United States Army Medical Department.[56] America's weather had finally become a national institution.

Throughout the early republican period, America's destiny was reflected in its weather. And the idea of *improving* the American habitat, its climate, and the fertility of its soil by cutting down forests, draining wetlands, and planting crops became one of the most important ideological justifications for the westward expansion of the United States. The new republic expanded westward under the sign of an environmental ideology, with the profound belief that its people, by virtue of their political system and their technological capacity, could transform North America. When, for instance, in the *Notes on the State of Virginia*, Jefferson suggests that the clearing of land has led to a westward expansion of warming winds, he is describing both a meteorological and a colonial phenomenon: "The Eastern and South-eastern breezes come on generally in the afternoon. They have advanced into the country very sensibly within the memory of people now living. . . . As the lands become more cleared, it is probable they will extend still further westward" (203). Americans did not simply reject colonial biogeography, they learned instead to employ it for their own expansionary purposes, reconceptualizing America from a colony to a nation grounded in the power of the American settlers to remake their environment. The advance of America, in short, could be felt in the westward advance of its moderating breezes, in the increasing salubrity of its air, and in the awakening fertility of its soil.

Just as eighteenth-century landscape gardeners built landscapes that would reflect the power and wealth of their patrons, Americans applied the ideology of improvement to an entire continent: America was "a landscape in the process of being made," one that expressed the powers of this newly emerging republic.[57] "Every acre, reclaimed from the wilderness," writes Adam Hodgson, "is a conquest of 'civilized man over uncivilized nature,' an addition to those resources which are to enable his country to stretch her moral empire to her geographical limits, and to diffuse over a vast continent the physical enjoyments, the social advantages, the political privileges, and the religious institutions, the extension of which is identified with all his visions of her future greatness."[58] Faced with the desertlike, short-grass drylands west of the Missouri, American settlers had little doubt that nature would soon cooperate with their needs. "Why may we not suppose," writes Josiah Gregg in 1844,

> that the genial influences of civilization—that extensive cultivation of the earth— might contribute to the multiplication of showers, as it certainly does of fountains. Or that the shady groves, as they advance upon the prairies, may have some effect upon the seasons. At least, many old settlers maintain that the droughts are becoming less oppressive in the West . . . then may we not hope that these sterile regions might yet be thus revived and fertilized, and their surface covered one day by flourishing settlements to the Rocky Mountains?[59]

By the 1870s, these ideas would be popularized in the science of Charles Dana Wilber, who argued that, historically, "Rain follows the Plough."[60]

The discourse on environmental change can be seen, then, as being among the most important of eighteenth- and nineteenth-century political discourses. Republicanism shaped American climate discourse. Words such as *liberty* and *freedom* thus had an ecological and biogeographical dimension because they were formulated within the context of reflection on the extent to which climate determined or was capable of being modified by good government. The prevailing idea, associated with Montesquieu, was that climate determined national character and social institutions. Other Enlightenment political thinkers, such as Hume and William Godwin, denied that climate had any real impact on political institutions. Godwin, for instance, notes the generally accepted view that climate is among those "physical causes which have commonly been supposed to oppose an immovable barrier to the political improvement of our species," rendering "the introduction of liberal principles upon this subject in some cases impossible." But he then argues that climate ultimately has little impact upon the progress of liberty, because of "the superior influence of circumstances, political and social."[61] Both sides of this debate agreed, however, that environments and governments mirrored each other, that bad weather was a sign of bad government, and that good government freed human beings from the boundary mechanism of an inferior nature, to assist nature in becoming her true self. And given the psychosexual and biogeographical component of this model, Americans were unable to distinguish between the power of technology and sexual power; technology was consistently identified with sexual capacity, production mapped on reproduction.

It was in France in 1785 that Jefferson first heard "the Nightingale in all it's perfection." He was less than impressed, writing that "I do not hesitate to pronounce that in America it would be deemed a bird of the third rank only, our mockingbird, and fox-coloured thrush being unquestionably superior to it" (*Papers* 8:241). Two years later, his impressions were more positive. In Petrarch's chateau Vaucluse, in Avignon, he experienced the "rich treat" of hearing "every tree and bush . . . filled with nightingales in full song." Writing to his daughter Martha, he requested that she seek to become acquainted with its music so that she will "be able to estimate it's merit in comparison with that of the mocking bird." Still, however, he remains a strong supporter of the American mockingbird whose song, unlike that of the nightingale, lasts "thro' a great part of the year" (*Papers* 11:369–70). On the same day, writing to William Short, he provided a more extended description of the experience: "This delicious bird gave me another rich treat at Vaucluse. Arriving there a little fatigued I sat down to repose myself at the fountain, which, in a retired hollow of the mountain, gushes out in a stream sufficient to turn 300 mills, the ruins of Petrarch's chateau perched on a rock 200 feet perpendicular over the fountain, and every tree and bush filled with nightingales in full chorus" (11:372). It was perhaps natural in such a place for Jefferson to reflect on the relationship between poetry, nightingales, and climate. The nightingales' "song is more varied, their tone fuller and stronger here than on the banks of the Seine," he writes, which helps explain "another circumstance, why there never was a poet North of the Alps, and why there never will be one. A poet is as much the creature of climate as an orange or palm tree." By 1787, Jefferson had no doubt about the capacity of the American climate to produce poets and birdsong, for he immediately suggested the idea of importing nightingales there: "What a bird the nightingale would be in the climates of America! We must colonize him thither." Like the canaries used to test the air quality of mines, singing nightingales, like rhyming poets, would be a sure indication of the salubrity of America's air and the temperate qualities of its climate.

Notes

The epigraph that opens this chapter is from "Dejection: An Ode," line 1, *The Poems of Samuel Taylor Coleridge*, ed. E. H. Coleridge (Oxford: Oxford UP, 1912), 362.

1. Silvio A. Bedini, *Thomas Jefferson: Statesman of Science* (New York: Macmillan, 1990), 73–75.

2. In a 1784 letter to James Madison, Jefferson provides detailed instructions about what was to be included in weather diaries (*Papers of Thomas Jefferson,* ed. Julian P. Boyd, 27 vols. [Princeton: Princeton UP, 1955–1997], 7:31).

3. Edwin T. Martin, *Thomas Jefferson: Scientist* (1952; New York: Collier, 1961), 115.

4. Bedini, *Thomas Jefferson,* 74.

5. Ibid., 1.

6. For the relationship between climate and colonial medicine, see my *Romanticism and Colonial Disease* (Baltimore: Johns Hopkins UP, 1999). For general discussions of eighteenth-century environmental medicine, see Ludmilla J. Jordanova, "Earth Science and Environmental Medicine: The Synthesis of the Late Enlightenment," *Images of the Earth: Essays in the History of the Environmental Sciences* (Lancaster: University of Lancaster, 1981), 119– 46; James C. Riley, *The Eighteenth-Century Campaign to Avoid Disease* (London: Macmillan, 1987); Frederick Sargent, *Hippocratic Heritage: A History of Ideas about Weather and Human Health* (New York: Pergamon, 1982); Alain Corbin, *The Foul and the Fragrant* (Cambridge, MA: Harvard UP, 1986); George Rosen, *From Medical Police to Social Medicine: Essays on the History of Health Care* (New York: Science History, 1974); Genevieve Miller, "'Airs, Waters, and Places' in History," *Journal of the History of Medicine and Allied Sciences* 17 (1962): 129–40. For medical meteorology in eighteenth- and early-nineteenth-century England and America, see James H. Cassedy, "Meteorology and Medicine in Colonial America: Beginnings of the Experimental Approach," *Journal of the History of Medicine and Allied Sciences* 24 (1969): 193–204; Gordon Manley, "The Weather and Diseases: Some Eighteenth-Century Contributions to Observational Meteorology," *Notes and Records of the Royal Society of London* 9 (1952): 300–07; Lieutenant-Colonel Edgar Erskine Hume, "The Foundation of American Meteorology by the United States Army Medical Department," *Bulletin of the History of Medicine* 8 (1940): 202–38. Stuart Peterfreund's chapter

(3) in this collection, on the role of weather observation in Gilbert White's *Natural History of Selborne*, provides a valuable perspective on the political dimensions of meteorological observation during this period.

7. *Writings of Thomas Jefferson, Being His Autobiography, Correspondence, Reports . . .*, ed. H. A. Washington, 9 vols. (New York: H. W. Derby, 1861), 4:159.

8. *Papers*, 2:195; emphasis added.

9. John C. Greene, *American Science in the Age of Jefferson* (Ames: Iowa State UP, 1984), 5.

10. *William Wordsworth*, ed. Stephen Gill (Oxford: Oxford UP, 1984), lines 24–26.

11. *Imagined Communities: Reflections on the Origin and Spread of Nationalism* (London: Verso, 1983).

12. Romantic discussions of weather have not received the attention they deserve. The best current treatment of the subject is Arden Reed's *Romantic Weather: The Climates of Coleridge and Baudelaire* (Providence: Brown UP, 1983). This study, though it is valuable in indicating the importance of the topic for Romantic writers, is seriously flawed by Reed's unsupported claim that the eighteenth century was not interested in weather because it could neither control nor measure it: "When Enlightenment thinkers seemed to lose interest, it was not because the weather became uninteresting; rather, their changing attitude may be understood in the context of the development of modern science. The new experimental science depended on exact, rigorous measurement of a wholly regulated environment, and this necessarily excluded the weather as a variable that could be controlled . . . in short, science moved indoors" (4). I would argue, in fact, the opposite: scientists during this period began to measure the weather by taking the meteorological instruments (the barometers, thermometers, hydrographers, and so on that first appeared during the seventeenth century) outdoors.

13. *Philosophy of History*, trans. J. Sibree (New York: Colonial, 1900), 80–81.

14. James Larson, "Not Without a Plan: Geography and Natural History in the Late Eighteenth Century," *Journal of the History of Biology* 19 (1986): 448. For the history of biogeography, see Janet Browne, *The Secular Ark: Studies in the History of Biogeography* (New Haven: Yale UP, 1983).

15. *Ecological Imperialism: The Biological Expansion of Europe, 900–1900* (Cambridge: Cambridge UP, 1986), 194.

16. For plant importation in seventeenth- and early-eighteenth-century England, see Douglas Chambers, *The Planters of the English Landscape Garden: Botany, Trees, and Georgics* (New Haven: Yale UP, 1993).

17. For a relevant account of this episode, see Tim Fulford, "Romanticism, the South Seas and the Fruits of Empire," *European Romantic Review* 11 (2000): 408–34.

18. *In the Wake of Cook: Exploration, Science, and Empire, 1780–1801* (London: Croom Helm, 1985), 14; see also his "A Presiding Genius of Exploration: Banks, Cook, and Empire, 1767–1805," *Captain James Cook and His Times*, ed. Robin Fisher and Hugh Johnston (Vancouver: Douglas and McIntyre, 1979), 21–39. For the role of English and French scientific institutions in furthering ecological control of the planet, see John Gascoigne, *Science in the Service of Empire: Joseph Banks, the British State and the Uses of Science in the Age of Revolution* (Cambridge: Cambridge UP, 1998); and Lucile H. Brockway, *Science and*

Colonial Expansion: The Role of the British Royal Botanic Gardens (New York: Academic, 1979); Michael A. Osborne, *Nature, the Exotic, and the Science of French Colonialism* (Bloomington: Indiana UP, 1994); and E. C. Spary, *Utopia's Garden: French Natural History from Old Regime to Revolution* (Chicago: U of Chicago P, 2000).

19. For Buffon and biogeography, see Gareth Nelson, "From Candolle to Croizat: Comments on the History of Biogeography," *Journal of the History of Biology* 11 (1978): 273–78.

20. Georges Louis Leclerc, Comte de Buffon, *Natural History, General and Particular*, trans. William Smellie, 9 vols. (London: W. Strahan and T. Cadell, 1785), 2:10. Cited parenthetically in the text hereafter as *NH.*

21. Hippocrates writes, for instance, "Heat is the foundation of all the functions in the body in the same way that it germinates the seed in earth, that it governs and regulates the Universe at large; it is the cause of all consumption and growth, visible and invisible: soul, reason, growth, movement, diminution, permutation, sleep, consciousness of all and everything, and never ceases to act" (*Hippocrates on Diet and Hygiene*, ed. John Precope [London: Zeno, 1952], 36).

22. *The Lay of the Land: Metaphor as Experience and History in American Life and Letters* (Chapel Hill: U of North Carolina P, 1975), 4.

23. *Letters of Alexander von Humboldt to Varnhagen von Ense*, trans. Friedrich Kapp (New York: Rudd and Carleton, 1860), 102.

24. See Thomas Laqueur, *Making Sex: Body and Gender from the Greeks to Freud* (Cambridge, MA: Harvard UP, 1990), 43–102.

25. *Collected Works*, ed. Arthur Friedman, 5 vols. (Oxford: Clarendon, 1966), 2:368.

26. *Don Juan*, in *Byron*, ed. Jerome J. McGann (Oxford: Oxford UP, 1986), 14:29–30.

27. *Natural History of Birds*, 3 vols. (London: A. Strahan and T. Cadell, 1792), 1:13.

28. For a discussion of Buffon's representation of American birdsong, see John Robert Moore, "Goldsmith's Degenerate Song-birds: An Eighteenth-Century Fallacy in Ornithology," *Isis* 34 (1943): 325.

29. *Natural History of Birds*, 1:16–17.

30. *Recherches philosophiques sur les Américains ou Mémoires intéressants pour servir à l'histoire de l'espèce humaine*, 2 vols. (London; Berlin: n.p., 1768), 1:1, 2:182.

31. *Philosophy of History*, 80–81; *Hegel's Philosophy of Nature*, ed. and trans. M. J. Petry, 2 vols. (London: Allen and Unwin, 1970), 2:80–83. Hegel is citing Johann Baptist Spix and Karl Friedrich Philipp Martius' *Travels in Brazil in the Years 1817–1820* (London: 1823–1831).

32. For the debate over the Old and New Worlds, see Martin, *Thomas Jefferson* 129–80; Gilbert Chinard, "Eighteenth Century Theories on America as a Human Habitat," *Proceedings of the American Philosophical Society* 91 (1947): 27–57; Antonello Gerbi, *The Dispute of the New World: The History of a Polemic, 1750–1900*, rev. and enl. ed., trans. Jeremy Moyle (Pittsburgh: U of Pittsburgh P, 1973); Henry W. Church, "Corneille de Pauw, and the Controversy over His *Recherches philosophiques sur les Américains*," *Publications of the Modern Language Association of America* 51 (1936): 178–207.

33. *Blake's Poetry and Designs*, ed. Mary Lynn Johnson and John E. Grant (New York: Norton, 1979).

34. See my *Romanticism and Colonial Disease*, 212–14.

35. *Queen Mab* 8.145–56, in *Shelley: Poetical Works*, ed. Thomas Hutchinson (London: Oxford UP, 1905).

36. Paula R. Feldman and Diana Scott-Kilvert, eds., *The Journals of Mary Shelley: 1814–1844* (Oxford: Clarendon, 1987), 1:174–76. She also requested that Shelley purchase a copy of Goldsmith's *Animated Nature* for her in October of that year; see Betty T. Bennett, *The Letters of Mary Wollstonecraft Shelley*, 3 vols. (Baltimore: Johns Hopkins UP, 1980), 1:53–54.

37. Mary Shelley, *Frankenstein: or, The Modern Prometheus. The 1818 Text*, ed. James Rieger (Chicago: U of Chicago P, 1974), 99.

38. Ibid., 105.

39. "The Deserted Village," *The Poems of Gray, Collins, and Goldsmith*, ed. Roger Lonsdale (London: Longman, 1969), lines 346–61.

40. *The Letters of John Keats, 1814–1821*, ed. Hyder Edward Rollins, 2 vols. (Cambridge, MA: Harvard UP, 1958), 1:397–98.

41. *Complete Poems*, ed. Jack Stillinger (Cambridge, MA: Harvard UP, 1982), ll. 30–43.

42. *Letters*, 2:243

43. *The Letters of Horace Walpole*, ed. Peter Cunningham, 9 vols. (London: Richard Bentley and Son, 1886), 7:365.

44. "Eighteenth Century Theories," 35.

45. *Ecological Revolutions: Nature, Gender, and Science in New England* (Chapel Hill: U of North Carolina P, 1989).

46. *Landscape and Ideology: The English Rustic Tradition, 1740–1860* (Berkeley and Los Angeles: U of California P, 1986), 14.

47. Martin, *Thomas Jefferson*, 126; *Papers of Thomas Jefferson* [hereafter *Papers*], 7:304.

48. *Writings of Thomas Jefferson*, 4:570.

49. Quoted in Martin, *Thomas Jefferson*, 126.

50. Writing to Ezra Stiles in 1784, Jefferson remarks: "I understand from different quarters that the Indians believe this animal still existing [*sic*] in the North and NorthWest tho' none of them pretend ever to have seen one" (*Papers* 7:305).

51. *Notes on the State of Virginia*, in *Thomas Jefferson: Writings* (New York: Library of America, 1984), 169, 176. Cited parenthetically in the text hereafter.

52. C. F. Volney, *View of the Climate and Soil of the United States of America* (London: J. Johnson, 1804), 266, 213–16.

53. *Writings*, 7:375.

54. *American Science*, 29.

55. *Writings*, 8:259.

56. For material on the development of the National Weather Bureau, see Hume, "The Foundation of American Meteorology"; and Alexander McAdie, "A Colonial Weather Service," *Popular Science Monthly* 44 (July 1894): 331–37.

57. Michael Williams, *Americans and their Forests: A Historical Geography* (Cambridge: Cambridge UP, 1988), 121.

58. *Letters from America Written during a Tour in the United States and Canada*, 2 vols. (London: Hurst, Robinson, 1824), 1:396–97.

59. *Commerce of the Prairies; or the Journal of a Santa Fé Trader*, 2 vols. (New York: H. G. Langley, 1844), 2:202–03.

60. Quoted in Henry Nash Smith, *Virgin Land: The American West as Symbol and Myth* (Cambridge, MA: Harvard UP, 1950), 211.

61. *Enquiry Concerning Political Justice* (Harmondsworth: Penguin, 1985), 146, 149

CHAPTER FIVE

Robinson Crusoe's Earthenware Pot

Science, Aesthetics, and the Metaphysics of True Porcelain

LYDIA H. LIU

Virginia Woolf once made a remarkable observation about Daniel Defoe's novel *Robinson Crusoe* (1719). Call it intuition or uncanny lucidity. Under her eyes, an insignificant detail, which has largely escaped the attention of Defoe's critics, emerges out of obscurity and becomes luminous all of a sudden. The illumination radiates from a plain earthenware pot that practically dominates the physical environment of Crusoe's world. Although Defoe's readers will remember that this pot is but one of many survival tools that Crusoe has invented during his solitary existence on the island, Woolf insists on seeing more. In her reading, the object acquires an enigmatic symbolism:

> Thus Defoe, by reiterating that nothing but a plain earthenware pot stands in the foreground, persuades us to see remote islands and the solitudes of the human soul. By believing fixedly in the solidity of the pot and its earthiness, he has subdued every other element to his design; he has roped the whole universe into harmony. And is there any reason, we ask as we shut the book, why the perspective that a plain earthenware pot exacts should not satisfy us as completely, once we grasp it, as man himself in all his sublimity standing against a background of broken mountains and tumbling oceans with stars flaming in the sky?[1]

Taking Crusoe's pot as a primary figure of representation in the novel, Woolf directs our attention to a productive metonymy between man and the thing he makes and to the possible limits of such metonymic figuring. In her reading, Defoe's (or rather Crusoe's) fixation on the solidity and earthiness of the pot takes on an aura of fetishism that evokes both the historicity of the metonymy and its aesthetic implications in the eighteenth century. The pot can thus be read as a fetish, though not a primitive's fetish but a modern man's, because it carries the symbolic burden of human intentionality that threatens to subdue the natural elements to his design. The image of Defoe or Crusoe roping the whole universe into harmony is just as disturbing as it is violent, which is, perhaps, what prompted Woolf to raise the rhetorical question toward the end of the quoted passage. But her question is not entirely rhetorical because it also casts a slight shade of ambiguity on Crusoe's dubious identity as the inventor and owner of the earthenware pot. Among other things, I attribute this ambiguity to the uncertain identity of the pot itself caused by the *accidental* happening of its making in the original context of Defoe's narrative. That which produces the accident of Crusoe's pottery, I argue, can be grasped both in terms of the circumstance of the novel's first publication in 1719 and in terms of the anachronism of Crusoe's mode of production, popularized by the classical political economists and criticized by Marx in *Capital*.[2]

From Science Fiction to Realism

Woolf's reading of *Robinson Crusoe* is intriguing for a number of reasons. I am particularly drawn to her suggestion of fetishism, and I wonder if we could further elaborate the figure of fetishized metonymy between Crusoe and his pottery not with a view to resolving Woolf's own ambiguity but with a view to following the traces of that metonymy to a larger, possibly global, network of metonymic exchange within which Defoe's earthenware episode was embedded and to which the novel *Robinson Crusoe* has made singular contributions. I emphasize the global network of metonymic exchange in this essay because we are dealing with some of the consequences of early modern global circulation that had predated and preconditioned European colonialism.

In a recently published study of early modern economy, Andre Gunder Frank again reminds us how, from the early fifteenth century through the beginning of the nineteenth century, "Europeans sought to muscle in on '*the richest trade in the world*,'" referring to the intra-Asian trade, and how colonization finally enabled Europeans to achieve that goal.

> Europeans derived *more profits* from their participation in the intra-Asian "country trade" than they did from their Asian imports into Europe, even though many of the latter in turn generated further profits for them as re-exports to Africa and the Americas. So the Europeans were able to profit from the much more productive and wealthy Asian economies by participating in the intra-Asian trade; and *that* in turn they were able to do ultimately only thanks to their American silver.[3]

Among the familiar Asian commodities sought after by the Europeans in early modern times were silk, tea, calicos, and porcelain (chinaware). Chinese porcelain, also known as true (white) porcelain, was traded around the globe and eagerly copied by potters elsewhere. By the late seventeenth and early eighteenth centuries, porcelain had become the single most fashionable luxury in the homes of the European aristocracy. Indeed, it was porcelain, not earthenware, that widely circulated in the global network of metonymic exchange in Defoe's time.[4] This is something we need to keep in mind when reexamining the earthenware episode in his novel.

The artifact Crusoe makes in *Robinson Crusoe* is called earthenware, *not porcelain*, and there seems no reason why we should ask the novel to do otherwise. However, Defoe's novel was written at the height of the European craze for true porcelain, and in the same period the author published several journalistic pieces arguing against imported chinaware and its negative impact on the British economy and morals.[5] Viewed against this background, the earthenware episode in *Robinson Crusoe* appears doubly interesting. Defoe's journalistic writing shows that he was not uninterested in the symbolic and technological difference between earthenware and porcelain. In fact, during King William's reign, Defoe himself attempted the manufacture of bricks and pantiles (S-shaped earthenware tiles developed in Holland) in response to the rising demand for construction materials for the rebuilding and expansion of London. In partnership with others, he became proprietor of a brickyard in the 1690s. According to Paula R. Backscheider, the Essex factory was Defoe's major business enterprise, from which he came to clear about six hundred pounds a year and which he believed to be a firmer foundation for his family's future economic security than his other projects.[6] But King William's reign was also a time when export Chinese porcelain prevailed on the global market and culminated in the rise of European chinoiserie, which would dictate the taste of the aristocracy in the next few decades. Deeply critical of this new trend, Defoe took King William and Queen Mary to task for having introduced four customs of excess that were imitated by the people and became worshiped by the whole kingdom: gardening, painting, East Indian calicos, and Chinese porcelain. Defoe argued that the royal taste had descended "into the humours of the common people so much, as to make them grievous to our trade, and ruining to our manufactures and the poor; so that the Parliament were oblig'd to make two Acts at several times to restrain, and at last prohibit the use of them." Of export porcelain, Defoe wrote:

> The queen brought in the custom or humour, as I may call it, of furnishing houses with china-ware, which increased to a strange degree afterwards, piling their china upon the tops of cabinets, scrutores, and every chymney-piece, to the tops of the ceilings, and even setting up shelves for their china-ware, where they wanted such places, till it became a grievance in the expence of it, and even injurious to their families and estates.[7]

Defoe's antipathy toward King William and Queen Mary aside, the curious configuration of ceramic objects in his life and writing is well worth pondering. As a business entrepreneur, he manufactured pantiles and earthenware, as does Crusoe in his novel, but at the same time Defoe was an outspoken critic of imported chinaware and went so far as to ridicule porcelain in the second volume of *Robinson Crusoe*. The motif of rivalry between earthenware and true porcelain in his writing no doubt expressed Defoe's protectionist stance against the penetration of the national market by foreign luxury products. Yet the economic rivalry was never a purely economic phenomenon but was readily translatable into metonymic associations at the discursive level, where earthenware would almost always evoke porcelain, and vice versa, in the eighteenth century. The presence of the earthenware pot in *Robinson Crusoe*, therefore, evokes porcelain by metonymic association and calls up the existence of the latter by virtue of its absence.

What I am trying to suggest here is that the rivalries among economies and civilizations in the eighteenth century seem to have undergone an extraordinary process of metamorphosis in Defoe's novel for it to become a tale of (white) man's *solitary survival in nature*. In that sense, Crusoe's experiment with earthenware is symptomatic of what I call the "poetics of colonial disavowal." The analysis that follows is an attempt to show that this poetics of colonial disavowal informs Defoe's storytelling in profound ways, both stylistically and contextually. To understand how it functions in the text, one cannot simply substitute one allegorical reading (of the survival tale) for another but would do well to interrogate the very absence of porcelain in the Crusoe episode and to try to explain how that absence conditions what the earthenware pot is doing, metonymically, in the novel. For porcelain is a significant absence in volume 1 of *Robinson Crusoe*. Like a ghost, the foreign object hovers over the borders of Defoe's writing and wrestles with the authorial hand that tries to exclude it from signification.

Even as the solidity and singularity of Crusoe's earthenware pot as recognized by Woolf seem ready to dissolve and metamorphose into something else, there is yet another level of complexity to be considered in this study. In keeping with the desire to control and regulate the value of luxury commodities on the inventory list of imported goods and to manufacture imitation products for competition on a global market, early-eighteenth-century Europe witnessed a growing scientific interest in how one might distinguish among true porcelain, soft-paste porcelain, and other types of ceramics, and how to fix those distinctions categorically. Alchemists and/or scientists put years of painstaking research into discovering and determining the differential scientific (interior) value of white porcelain as opposed to the more familiar European soft-paste (*pâte tendre*) ware and earthenware. My research shows that the meaning of true porcelain at this time was shot through with Europeans' curiosity about the basic components of chinaware, then reported to be the Chinese clay kaolin and the porcelain stone petuntse. The quest for the local variants of these materials, therefore, introduced an interesting metaphysical disjuncture between true porcelain on the one hand and earthenware and soft-paste porcelain such as faience and delft ware on the other. In other words, chinaware was singled out to represent essential difference from ordinary ceramics and soft-paste porcelain from a scientific point of view because that difference and the distinction it conferred on the object mattered a great deal in terms of quality and value to contemporary merchants, scientists, collectors, and manufacturers.[8]

That said, what new insight can we glean from Crusoe's earthenware episode? Could it lead to a new interpretation of *Robinson Crusoe*? In a fine chapter, "Robinson Crusoe and Friday," in his study *Colonial Encounters*, Peter Hulme takes issue with two main strands of criticism of Defoe's novel since Ian Watt's influential 1957 book *The Rise of the Novel*, namely, Watt's own thesis of economic individualism and "formal realism" on the one hand and the "'spiritual' reading" of the novel as "a Puritan fable" on the other.[9] With regard to *Robinson Crusoe*'s place in literary history as the first true work of realism, Hulme identifies the following grounds as central to Watt's argument. First, *Robinson Crusoe* is the preeminent novel of the "individualism" that characterizes modern realist fiction; second, it fulfills more generally the criteria of Watt's formal realism; and, finally, the novel demonstrates, in Crusoe's wanderings, "the dynamic tendency of capitalism itself, whose aim is never merely to maintain the status quo, but to transform it incessantly."[10] Hulme argues that these grounds fall apart when the episodic or seemingly formless plot of Defoe's novel fails to live up to Watt's own criteria of formal realism and intrinsic coherence—that touchstone of bourgeois aesthetics.

To help solve this difficulty and recuperate the formal sophistication and narrative coherence of *Robinson Crusoe*, some critics began to approach it as a Puritan spiritual tale by analogy with the Puritan journal.[11] According to this line of argument, there is an underlying spiritual pattern in *Robinson Crusoe* that enables the narrator to make sense of his daily experience by negotiating providential meanings in an immediate recording of the crowded sensations of the lived moment. Rather than formal realism, the spiritual pattern is what gives Defoe's narrative its true significance. However, this reading strategy and its pursuit of narrative coherence run up against another set of interpretive problems. In Hulme's words, "the 'spiritual' reading of *Robinson Crusoe* attempts—unsuccessfully—to remedy the scandal of the secular text whose interpretation is not guided by any authorial voice, but which has been published as the character's own story, 'Written by Himself,' an assertion, as Watt rightly saw, of the primacy of individual experience as defiant in its own—fictional—way as Descartes' *cogito ergo sum*" (*CE* 179). In a nutshell, the disagreement between the realist and spiritual readings of *Robinson Crusoe* has led to two very different Defoes by Hulme's account: against Watt's "modern" Defoe—Defoe/Richardson/Fielding—is set a seventeenth-century Defoe—Milton/Bunyan/Defoe.

Seeking a possible alternative to this, Hulme proposes to read *Robinson Crusoe* as a Caribbean book or colonial narrative. He argues that the new subjectivity that emerges in the course of Defoe's novel "is simultaneously an individual and a national consciousness, both forged in the smithy of a Caribbean that is—as of course the Caribbean still is to England—both parabolic and historical at the same time. Concomitantly, the social relationships involved are simultaneously personal and international" (*CE* 216). The same view was forcibly expressed by James Joyce, whose words are quoted by Hulme: "The true symbol of the British conquest is in Robinson Crusoe. . . . The whole Anglo-Saxon spirit is in Crusoe; the manly independence and the unconscious cruelty; the persistence; the slow yet efficient intelligence; the sexual apathy; the practical, well-balanced, religiousness; the calculating taciturnity."[12] By emphasizing the social relationship of the personal and international, Hulme seems to be shifting the ground of Defoe's realism from the earlier critics' formalist argument and spiritual reading to a study of colonial subjectivity. His analysis of the colonial exchange between Crusoe and Friday, for example, makes a good case toward illustrating that "the imperial production of *Robinson Crusoe* as a boys' [*sic*] adventure in the nineteenth century inevitably foregrounds the colonial alibi—the man alone, on a desert island, constructing a simple and moral economy which becomes the basis of a commonwealth presided over by a benevolent sovereign" (*CE* 222).

Although Hulme's insight is immensely useful for our rethinking of Defoe's novel and English literature in general, the New Historicist twist on realism in his argument still begs the question of how colonialism grounded and governed its own realism. Indeed, the imperial production of *Robinson Crusoe* as a boy's adventure does produce a colonial alibi, but the realism of that alibi needs further interrogation. In *Emile*, for instance, Rousseau provides what has long been regarded as the standard pedagogical reading of Defoe's novel as a children's tale. In Book 3, Rousseau writes:

> Robinson Crusoé dans son île, seul, dépourvu de l'assistance de ses semblables et des instrumens de tous les arts, pourvoyant cependant à sa subsistance, à sa conservation, et se procurant même une sorte de bien-être, voilà un objet intéressant pour tout âge, et qu'on a mille moyens de rendre agréable aux enfans. *Voilà comment nous réalisons l'île déserte qui me servoit d'abord de comparaison.*

> [Robinson Crusoe on his island, deprived of the help of his fellow-men, without the means of carrying on the various arts, yet finding food, preserving his life, and procuring a certain amount of comfort; this is the thing to interest people of all ages, and it can be made attractive to children in all sorts of ways. *We shall thus make a reality of that desert island which formerly served as an illustration.*][13]

What does Rousseau mean by "voilà comment nous réalisons l'île déserte qui me servoit d'abord de comparaison"? Perhaps this is where one might hope to gain some insight into the status of the real both in and outside Defoe's novel. The matter is further complicated by Rousseau's observation that Émile's mimicking of Robinson Crusoe is like building a *"vrai château en Espagne* de cet heureux âge, où l'on ne connoît d'autre bonheur que le nécessaire et la liberté [*genuine castle in the air* of this happy age, when the child knows no other happiness but food and freedom]" (*EO* 156; *E* 148; emphasis added). The interesting slippage between "nous réalisons" and "vrai château en Espagne" in Rousseau's language could be explained away by the assumption that a child's unfettered imagination cannot distinguish between these things in the first place. But *Émile* is a philosophical treatise written for educated adults whose claims to reason were predicated precisely on their ability to distinguish between the realm of the real and castles in the air and who would say of *Émile*, as Rousseau himself puts it in his preface, that "on croira moins lire un traité d'éducation que les rêveries d'un visionnaire sur l'éducation [this is not so much a treatise on education as the *visions of a dreamer* with regard to education]" (*EO* 2; *E* 2; emphasis added). Rousseau's tacit admission to having dreamed up an elaborate castle in the air suggests an interesting figurative exchange between *Émile* and *Robinson Crusoe*. I argue that this economy of exchange took place in the imaginary realm of proto-science fiction, where Rousseau's technology of pedagogy found itself responding creatively to elements of science fiction in Defoe's novel.

Which is to say that *Émile* foregrounds the science fiction of Defoe's novel by casting itself as a science fiction of sorts—one about the technology of pedagogy— although Rousseau was neither the first nor the last to do so. Jules Verne, the celebrated inventor of science fiction, inherited this economy of figurative exchange from Defoe and Rousseau in the nineteenth century and openly acknowledged *Robinson Crusoe* as a major source of inspiration for all his works. "My story," says Captain Grant of *The Children of Captain Grant* (1868), "is that of all Robinsons thrown up on an island."[14] I should also mention that recent cinematic reincarnations of Defoe's hero, such as *Robinson Crusoe on Mars* and *Enemy Mine*, further attest to a living tradition of science fiction in the twentieth century that shows no sign of letting go of the Verne paraphernalia.[15] The curious circularity among Defoe, Verne, and some of today's science fiction says something very interesting about the historical status of the earlier text as an anticipation, a form-giving moment calling for its actualization in repeated imitations by future writers. Pierre Macherey has made a brilliant observation about the meaning of this circularity in an essay on Jules Verne: "All those who have wanted to settle their accounts with *Robinson Crusoe*—Jules Verne amongst others—have achieved this in so far as it allowed them to criticize a certain representation of origins. This critique depends on revealing the *circularity of origins*."[16] For Verne, each rewriting of Crusoe constitutes a new statement about science fiction and its reworking of the terms of reality: "To outline a new Crusoe, intended as a symbol of reality, is to pose the problem of fiction and its reality, and consequently the problem of its mistaken irreality: all these problems in a single necessary moment."[17]

My point is not to introduce a deliberate rupture between *Robinson Crusoe* and Watt's realist tradition in order to reclassify Defoe's work as science fiction, although that certainly is a genuine possibility. For the purpose of this essay, I am more interested in pursuing what there is about Defoe's novel to which Rousseau, Verne, and many others have responded so spontaneously and powerfully through the centuries, in spite of its worn motif of maritime adventure. This analysis, in turn, may help us rethink the realism of *Robinson Crusoe* as *a process of becoming*, whereby one must learn to overlook the traces of science fiction a priori in order to imagine the work as a realist novel. This learning process finds its own historical roots in the poetics of colonial disavowal I have mentioned—shared by Defoe and the majority of his critics—that almost always grounds the realism of this novel in the purported absence of the other and, therefore, at the expense of the other.

Let me be more specific. The science fiction of Defoe's time, like that of our own, was closely linked to what was going on in the actual laboratory. In regard to true porcelain, serious laboratory experiments were undertaken by Ehrenfried Walther von Tschirnhaus and the alchemist Johann Friedrich Böttger in Saxony at the turn of the century, leading to the founding of the Meissen factory in 1710. We know that between 1712 and 1722 the Jesuit missionary Père d'Entrecolles (whose Chinese name is Yin Hongxu) conducted his famous industrial espionage on the Chinese porcelain metropolis Jingdezhen (King-te-Tching) and sent back two long letters from China. These letters were subsequently excerpted extensively by Jean-Baptiste Du Halde in his book *Description géographique, historique, chronologique, politique, et physique de l'empire de la Chine* (1735) and became popular across Europe as a result. (My speculation: Defoe could have read or heard about d'Entrecolles's first letter of September 1, 1712, which was published as early as 1716 in the *Journal des Savants*). D'Entrecolles's report on the preparation of the essential components of porcelain—kaolin and petuntse—aroused widespread scientific interest among European scientists.[18] The French physicist René Antoine Ferchault de Réaumur subsequently used the information d'Entrecolles gathered to carry out his much-publicized research on the Chinese stone as he attempted to transmute glass into true porcelain. Britain was very much a part of this process in the development of its own scientific and aesthetic apparatus.[19] When Defoe published the first two volumes of *Robinson Crusoe* in 1719, Britain had not yet discovered the secret of white porcelain and relied on trade and its growing maritime power to meet the domestic demand for imported Chinese and Japanese trade porcelain and then Meissen products. Not until after the middle of the eighteenth century did Britain begin to acquire large-scale manufacturing capabilities—first in soft-paste porcelain and then in true porcelain.

Defoe's manufacture of bricks and pantiles was in every sense a part of this larger story. However, a biographical reading of *Robinson Crusoe* would not satisfy us as a way of anchoring its ceramic realism because the novel also participated in a larger, collective enterprise that enabled the scientists and novelists to *fantasize* collectively about a goal and an object in the manner of science fiction. Regardless of how successful Defoe was with his pantiles in real-life manufacture, what Robinson Crusoe does so well in the novel is mimic the experiments of Tschirnhaus, Böttger, Réaumur, and Dutch potters who dreamed of being the first to replicate porcelain, especially white porcelain.[20] Even though the scientists unabashedly relied on industrial espionage or stolen specimens brought to Europe by sea merchants, *Crusoe's solitary experiment requires no external help*. Was porcelain not *a type of earthenware* that a British man could have invented all by himself?

Indeed, the author of *Robinson Crusoe* maneuvers the figural rivalry of earthenware and porcelain of his time so skillfully that one could easily overlook the traces of science fiction in the novel, as they have been consistently overlooked by his critics in the past and present. I argue that such maneuvering anticipated and contributed to a historical process in which the elements of science fiction seem to have fallen out of the picture altogether so that a realist or spiritual reading would come to dominate the interpretation of the novel, often conceived sans volumes 2 and 3. The same process has coincided with a historical development whereby Europe's increasing mastery of the technologies of other civilizations produced the very ground on which the primitiveness and backwardness of those civilizations would be mythologized.

What a Translated Text Can Tell Us That the Original Cannot

Let us recall one or two important details in the earthenware episode of *Robinson Crusoe*. Crusoe's experiment with pottery making takes place after he has successfully learned the rudimentary techniques of agriculture on the island. He has spent two solitary years on the island and has begun to harvest the fruits of his labor, such as barley, corn, and rice. One day, realizing that he needs some containers for preparing corn and grains and so forth, he sets out to make earthen vessels. It takes him two months to find the clay, dig it, temper it, bring it home, and work it into reasonable shapes. In the course of doing this, he has to overcome numerous difficulties trying to raise the clay from the ground and to prevent its cracking under the violent heat of the sun. After repeated failures and frustrations, Crusoe succeeds in making two large, sunbaked earthenware vessels, as well as some little round pots, flat dishes, pitchers, and pipkins (small earthen pots) (Figure 5.1). Although Crusoe is not completely happy with the large pots and calls them "two large earthen ugly things," he finds them useful. He puts them in two big wicker baskets and stuffs the space between the pots and baskets with dry rice and barley straw, intending to use these to hold dry corn and grains. The problem is that his sunbaked vessels can neither hold water nor withstand fire. Crusoe being Crusoe, he

Robinson ayant besoin de Vases pour tenir des liquides essaye de faire des pots et autres Utensiles avec de l'argile.

Figure 5.1. François Aimé Louis Dumoulin, *Collection de cent-cinquante gravures représentant et formant une suite non-interrompue des voyages et aventures surprenantes de Robinson Crusoé*, 2nd ed. (Vevey: Imprimerie de Loertscher et Fils, 1818), Pl. 45: Robinson ayant besoin de vases. Engraving. Courtesy of the University of California, Berkeley, Library.

never runs out of luck. Instead he makes a chance discovery of some sort of porcelain, or what some would have called chinaware:

> It happen'd after some time, making a pretty large Fire for cooking my Meat, when I went to put it out after I had done with it, I *found a broken Piece of one of my Earthen-ware Vessels in the Fire, burnt as hard as a Stone, and red as a Tile.* I was agreeably surpris'd to see it, and said to my self, that certainly they might be made to burn whole if they would burn broken.
>
> This set me to studying how to order my Fire, so as to make it burn me some Pots. I had no Notion of Kiln, such as the Potters burn in, or of glazing them with Lead, tho' I had some Lead to do it with; but I plac'd three large Pipkins, and two or three Pots in a Pile one upon another, and plac'd my Fire-wood all round it with a great Heap of Embers under them; I ply'd the Fire with fresh Fuel round the out-side, and upon the top, till I saw the Pots in the inside red hot quite thro', and observ'd that they did not crack at all; when I saw them clear red, I let them stand in that Heat about 5 or 6 hours, till I found one of them, tho' it did not crack, did melt or run, for the Sand which was mixed with the Clay melted by the violence of the Heat, and would have run into Glass if I had gone on; so I slack'd my Fire gradually till the Pots began to abate of the red Colour, and watching them all Night, that I might not let the Fire abate too fast, in the Morning I had three very good, I will not say handsome Pipkins; and two other *Earthen Pots,* and as hard burnt as cou'd be desir'd; and one of them perfectly glaz'd with the Running of the Sand.[21]

Crusoe's experiment has the innocent appearance of a chance discovery and, at first glance, does not stand out from the rest of his ingenious inventions at all. Nor does Crusoe make a verbal distinction between his sunbaked pottery and the new invention as he continues to call the perfectly glazed vessel "earthen pot," as if the firing process made no qualitative difference whatsoever (Figure 5.2).[22] Nevertheless, the similes I highlighted—"burnt as hard as a Stone, and red as a Tile"—seem to gesture toward something without actually naming it. The figurative equivalence running through earthenware, stone, and tile invites us to see the earthenware in a different light. Suppose we allow the subjunctive element in the similes to play itself out a bit. Suppose Crusoe's similes behaved like a slip of the tongue. Would they suggest traces of a different order of meaning, which Defoe has banished perfunctorily from the narrative but which has, somehow, found its way back into the text through the similes? Indeed, why "hard as a stone, and red as a tile"?

The short answer is that Crusoe's earthenware pot simultaneously evokes and disavows the palpable presence of Chinese porcelain and all the associative meanings this image could call up in the minds of Defoe's contemporary European readers, like chinaware, delftware, faience, maiolica, the porcelain towers that were being constructed in parks and gardens around Europe, chinoiserie, and so on. Now, a more interesting question is how we unravel the (con)textual implications of this simultaneous act of evocation and disavowal, and on what grounds. Before

Robinson aprés plusieurs essays infructueux réussit à cuire ses Vases de terre en les environnant de feux bien ardents.

Figure 5.2. F.A.L. Dumoulin, *Collection de cent-cinquante gravures*, Pl. 46: Robinson après plusieurs essays. Engraving. Courtesy of the University of California, Berkeley, Library.

I elaborate on the similes themselves in the European context, I would like to bring a heretofore unknown text to the attention of Defoe scholars, namely, a foreign-language translation of *Robinson Crusoe*. This text is fascinating in the sense that it offers itself up as a belated *metonymic reminder* of the traces of porcelain making in the original novel by renaming the object as such. The work I have in mind is the first Chinese translation of *Robinson Crusoe* by the celebrated translator Lin Shu and his English-language informant Zeng Zonggong, who published the first two volumes of the novel in classical Chinese in 1905–1906.[23] Lin and Zeng's work introduces what I call a "metonymic encounter" between Defoe's novel and its Chinese translation. The surprise encounter has the effect of estranging the reader, à la Viktor Shlovsky, from an otherwise thoroughly familiar English text.[24]

Let us consider the translated scene of Crusoe's experiment with clay and firing. In the Chinese version, Lin and Zeng took the liberty of correcting the novel's fuzzy terminology of earthenware, not suspecting that Crusoe's work could well have been a piece of *science fiction* in Defoe's time. Where the author uses phrases such as *Earthen Ware*,[25] *Earthen Pots*, *Earthen-ware vessels*, and so forth indiscriminately to designate the products of Crusoe's labor, Lin and Zeng decided to alter them to better reflect what seems to them the technological progress Crusoe makes from an elementary form of pottery: *wa*, to porcelain, *ci*, or *tao*.[26] (Incidentally, the one place where Defoe does speak of china and porcelain is in his description of Crusoe's travels in China in the second volume of the novel. I will tackle this episode later.) The Chinese translators' decision to "improve" the original text opens up an interesting interpretive space in which the historicity of the original stands exposed and is held accountable to the translated text.

This seems to reverse the usual relationship between the original text and its translation, as the burden of translatability shifts the plenitude of meaning onto the act of translation or interpretation. In that sense, I find Walter Benjamin's evocation of what seems to be a ceramic object in "The Task of the Translator" extraordinary, if not utterly surreal:

> Fragments of a vessel which are to be glued together must match one another in the smallest details, although they need not be like one another. In the same way a translation, instead of resembling the meaning of the original, must lovingly and in detail incorporate the original's mode of signification, thus making both the original and the translation recognizable as fragments of a greater language, just as fragments are part of a vessel. For this very reason translation must in large measure refrain from wanting to communicate something, from rendering the sense, and in this the original is important to it only insofar as it has already relieved the translator and his translation of the effort of assembling and expressing what is to be conveyed.[27]

Although *Gefäß* (vessel) need not refer to a ceramic object, the word *Scherben* (shards or fragments) strongly suggests that Benjamin probably did have that image in mind.[28] If the originality of his theory of translation turns on the figure of fragmented shards glued together to form a whole vessel, what are we supposed to make of the figure itself?[29] Lin and Zeng's rendering of Crusoe's earthenware pot certainly does not resemble the meaning of the original, nor do they seem to incorporate the original's mode of signification in the manner of matching the shards of a vessel. Since the earthenware pot itself now becomes the object of interpretation, the problem of meaning Lin and Zeng's translation raises for us is of a different order and is, perhaps, more interesting than the challenge of the pure language (*reine Sprache*) that Benjamin poses to translators and theorists.[30] What I mean is that the literality of Benjamin's ceramic figure must be interrogated in the performativity of translation, above and beyond whatever analogous value it might still carry *for translation.*

As a central trope of translation, the figure of the vessel can, therefore, be tested by Lin and Zeng's peculiar rendering of Crusoe's earthenware from English to classical Chinese. Contrary to the enormous symbolic significance Woolf once attached to this humble object, Lin and Zeng stare Crusoe's earthenware pot in the face and try to pin down the *literal* signification of Crusoe's experiment in the classical Chinese terminology with which they were familiar. The solidity and earthiness of the pot thus turn out to be very different degrees of solidity and earthiness as we follow Crusoe making progress step by step from the sunbaked *wa* to a chance discovery of *ci* or *tao*.

What Lin and Zeng did to Defoe's novel seems to push the literal meaning of Crusoe's earthenware pot to the limit by referring it *elsewhere.* They took Defoe's choice of words and transformed it into an itinerant sign in search of its classical Chinese equivalent that had been established through centuries of commercial usage. The words *ci* and *tao* are, therefore, chosen as the closest equivalent to what Crusoe claims to have discovered, which suggests that the translators are retranslating something back to what has always already been translated as "porcelain" and "chinaware" by Europeans in the past but has been overshadowed by Defoe's celebration of a British man's ingenuity in the novel. D'Entrecolles, for instance, glossed the meaning of *porcelaine* specifically as *tseki,* a French romanization of the characters *ci* and *qi* (literally, "porcelain vessel") of the time. In the first letter about Jingdezhen he sent back to Europe in 1712, d'Entrecolles wrote: "La porcelaine s'appelle communément à la Chine *tseki.*"[31] Although unaware of the discursive history of porcelain making between China and Europe that nonetheless preconditioned their own efforts as translators, Lin and Zeng's work effectively interrogates the *literality* of Defoe's similes and turns them inside out in the Chinese text.

In the Chinese version of the earthenware episode, the similes of stone and tile disappear and are replaced by the word *ci* as the translators rename some of Crusoe's "earthen pots" while preserving *wa* and *waqi* for his sunbaked pottery.[32] Where

Crusoe says "I found a broken Piece of one of my Earthen-ware Vessels in the Fire, burnt as hard as a Stone, and red as a Tile," the classical Chinese text reads (when retranslated into English): "In the ashes I found a broken piece of the clay that, having been fired so long, had turned into a porcelain tile [*ciwa*]."[33] By renaming the object "porcelain," the Chinese text renders entirely visible the direct material ties that had existed between China and Europe since the sixteenth century. That is to say, Crusoe's improved "pottery" is no more plain earthenware than Tschirnhaus's, Böttger's, and other Europeans' experimental replicas of Chinese porcelain. On the contrary, these belonged to an era when Europe was modernizing itself in the arts, science, technology, and material culture and did so by colonizing, appropriating, and (epistemologically) primitivizing the other civilizations.[34]

In the early eighteenth century, Chinese *ci* or *tao* was a *desired* equivalent of the European "porcelain" or "chinaware." Imitations of Chinese Yixing ware, known as red stoneware in Europe, and white porcelain may be said to represent a material act of translation aimed at overcoming any inaccuracies or nonequivalences of the sign with respect to the material it designated.[35] Their goal was to be able to reproduce the exact and equivalent value of true porcelain in terms of what was then called *Porzellan* or chinaware, which explains why the distinctions between porcelain, faience, and earthenware became of paramount importance.[36] Curiously, Defoe's intervention consisted in severing that equivalence and decontextualizing the transcultural meanings of Crusoe's labor to turn it into a myth. However, the myth is not complete and has left some traces behind because, as I have suggested, Crusoe's mimicking of the contemporary experiment inadvertently articulates his own condition of *science fiction*.

Being a business entrepreneur himself in brick and pantile manufacture, Defoe knew enough of the subject to describe Crusoe's experiment in the common parlance of his time, that is, as stone or tile. His description of the color of the object instantly reminds us of the red stoneware (*Jaspisporzellan*) and red Dutch floor tiles that John and David Elers made in England and Tschirnhaus and Böttger were producing after 1708 at the factory in Dresden–Neustadt, both trying to imitate Chinese Yixing ware through the example of Dutch potters.[37] But I am not trying to get at a realistic or referential reading of Defoe's novel, nor do I wish to take the posed naiveté of the narrator literally. There is no reason why Crusoe's fictional naiveté should be an alibi for us not to explore what *alternative reading* of Defoe's novel could perhaps emerge from the set of circumstances I am putting together. I argue, rather, that Defoe's use of the similes of stone and tile is more substantial than a mere rhetorical turn of language. Like the Chinese translation, the similes gesture toward what has been disavowed in the novel.

The disavowal is singularly striking when we consider further how Defoe discusses Chinese porcelain in some other venues. In an account of Queen Mary's passion for chinaware in *From London to Land's End*, for example, Defoe exhibits up-to-date knowledge of the differences between chinaware and soft-paste porcelain.

> Her Majesty had here a fine apartment, with a set of lodgings for her private re-treat only, but most exquisitely furnished, particularly a fine chintz bed, then a great curiosity; another of her own work while in Holland, very magnificent, and several others; and here was also her Majesty's fine collection of *Delft ware*, which indeed was very large and fine; and here was also a vast stock of fine *china ware*, the like whereof was not then to be seen in England; the long gallery, as above, was filled with this china, and every other place where it could be placed with advantage.[38]

Delft ware is known to be a type of soft-paste European porcelain, and the term is often used interchangeably with "faience" due to the regional differences of European porcelain manufacture in the early eighteenth century. George Savage restricts the proper use of "faience" to "pottery having a glaze made opaque with tin oxide *when it has been made in Germany or France*. The same kind of pottery made in Italy is termed *maiolica*, and in Holland and England, *delft*. The geographical line of demarcation is a little difficult to draw rigidly. It would, in fact, be better to regard the definitions of *maiolica* and *delft* as fixed, and to refer to tin-glazed pottery made elsewhere as *faïence*."[39] Whether or not we accept Savage's definition of early European soft-paste imitations of Chinese porcelain, this category was in the process of being *conceptually* differentiated from earthenware, on the one hand, and imported chinaware, on the other. Could we surmise that Defoe's knowledge of the ceramics, *not the lack of such knowledge*, was the likely source of his disavowal?

The question draws our attention to a possible process of substitution in the figuring of Crusoe's earthenware pot, which one might call a metaphorical endeavor. Defoe's substitution of earthenware for porcelain seems to contradict the metonymic impulse of the similes, which unwittingly evoke that which is being disavowed. The conflicting coexistence of the metaphorical and the metonymic in the figuring of the earthenware pot is what enables the simultaneous disavowal and evocation of porcelain in *Robinson Crusoe*. As readers may recall, Woolf sees a plain "earthenware" pot in the foreground and tries to wrestle symbolic meanings from it, but she does so without taking Defoe's metaphorical disavowal of porcelain into consideration. In the very act of questioning the self-made British hero, her reading has the effect of reproducing the mirage of a self-sufficient British man who lords it over the rest of the world as if it were, indeed, "a background of broken mountains and tumbling oceans with stars flaming in the sky," strangely devoid of civilizational resources and passively submitting to the manipulation of the white man's technology. But has Britain or Europe been self-sufficient and self-explanatory in the arts, science, and technology? Of course not. As my reading of the earthenware episode has demonstrated, the idea of self-sufficiency is itself the ideological effect of a storytelling (and a Defoe industry) that aims to produce its own realism by selectively disavowing the elements of science fiction as a condition of the novel's realism.

At the heart of it all lies the poetics of colonial disavowal. The earthenware episode is by no means an isolated example. Hulme, for example, mentions two other such instances of disavowal, both having to do with Crusoe's pedagogical efforts to civilize Friday, the cannibal. The key episodes in Friday's education center on the two aspects of Carib technology—the barbecue and the canoe—that Europe learned from the Caribbean. In the novel, however, Crusoe teaches Friday how to barbecue and how to build a canoe, remarking that Friday is filled with gratitude and admiration for his technology. "The 'ignorance' of the savage Caribs is *produced by* the text of *Robinson Crusoe*," observes Hulme, "which enacts a denial of those very aspects of Carib culture from which Europe had learned" (*CE* 210–11).

The Metaphysical Turn of True Porcelain

The concept of true porcelain was a European coinage with no corresponding term or category in native Chinese discourse on porcelain.[40] Eighteenth-century European potters and scientists had invented this concept to distinguish between Chinese porcelain and its European counterpart. The introduction of true porcelain thus instituted a metaphysical divide between real porcelain and that which merely looked real. In this scheme of things, true porcelain specifically referred to Chinese porcelain, whereas European faience was relegated to the category of the superficial, waiting to be improved. This seems to be mapping true porcelain onto the metaphysics of eighteenth-century science in which truth was supposed to reside inside, always underneath the surface. But how does one go about unmasking the truth of porcelain without breaking it? What would be the potential scientific and economic return of such unmasking?

The painstaking quest for the right clay—kaolin—and the right stone—petuntse—in the eighteenth century is illuminating because it brings our discussion of the metaphysics of true porcelain to the scientific realm of value and truth. From the time when Europeans believed that true porcelain was made by burying an earthenware piece in the ground for a long time to their discovery of local variants of kaolin and petuntse, there had been a fascinating history of cross-cultural value making that had everything to do with the assumed or contested translatability of the sign, knowledge, and epistemology between China and Europe.

European writers and experimentalists had long been interested in the mystery of the precious translucent material of Chinese porcelain, and with the exception of a few isolated experiments, their speculations had hardly reached beyond Marco Polo's early account. Consider the following passage from a 1617 publication:

> A large mass of material composed of plaster, egg- and oyster-shell, of sea-locusts and similar creatures, is well mixed until it is of one consistency. It is then buried by the head of the family, who reveals the hiding place to only one of his sons. It must remain in the ground for eighty years without seeing the light of day. After this time has elapsed, the heirs must take it from the ground, and use it to make the beautiful translucent vases of such perfect form and colour that no critic could find fault with them.[41]

Variants of the same myth survived through the seventeenth and early eighteenth centuries and found their way into the works of such writers as Sir Thomas Browne, Francis Quarles, William Cartwright, Richard Knolles, Thomas Shadwell, Sir Frances Bacon, John Donne, and Alexander Pope. William Forbes has pointed out that "Bacon writes in *Sylva Sylvarum* (1627) of 'Porcellane, which is an Artificiall Cement buried in the Earth a long time' and in *New Atlantis* (1627) of 'several Earths, where we put divers Cements, as the Chinese, do their Porcellane.'" He also points out that Donne includes these lines in his "Elegie on the Lady Marckham": "As men of China, after an ages stay,/Do take up Porcelane, where they buried Clay." Forbes further notes that "China's Earth" is evoked precisely in this meaning by Pope in his poem *The Rape of the Lock*.[42]

Defoe did not succumb to the old myth and knew what he was talking about when he made his fictional hero discourse on the clay, kiln, glaze, lead, firing, sand, glass, stone, tile, and so on. Having owned a brick factory himself, Defoe's knowledge of contemporary experiments with porcelain, faience, and delftware was sufficiently up to date. Prior to his Essex factory, several attempts had been made to reproduce the Dutch version of Chinese Yixing ware using the local clay found in England. Francis Place (1647–1728) undertook the experiments with gray stoneware in his workshop at Manor House, York. His work was followed by that of John Dwight (1637–1703), who started a pottery in Fulham and was granted a patent to make porcelain in 1671. According to one account, Dwight brought a lawsuit against the Elers brothers for infringement of his patent. The Elers arrived from Holland before 1686, working in Hammersmith at first and later starting a factory in Staffordshire. They made red unglazed stoneware in the manner of Chinese Yixing ware. The lawsuit showed that the ware produced by Dwight and by the Elers was identical.[43] The Dutch connection in Defoe's manufactory was also strong.[44] According to Backscheider, Defoe's knowledge of pantiles and of civet suggests that "he may have gone to Holland for a short business trip at least, but the numbers of Dissenters educated there and of English merchants who regularly visited the continent confuse the issue. Even had he been in Holland, Defoe probably would have had to hire a Dutch foreman or overseer." As late as 1726, Defoe's trade in pantiles made him a pioneer in manufacturing what was still called "a late Invention in *England.*"[45]

While this was going on, Tschirnhaus and Böttger were developing their own red stoneware with the help of a Dutch potter in Dresden–Neustadt (before the formal establishment of the Royal Saxon Porcelain Manufactory in Meissen on January 23, 1710). Their goal, however, was to replicate white porcelain.[46] Both men were Defoe's contemporaries and were in the service of Augustus II (Augustus the Strong), Elector of Saxony and King of Poland (1670–1733), whose passion for Chinese porcelain is legendary. Augustus is said to have exchanged a regiment of dragoons for forty-eight Chinese vases. In his lifetime, he owned thousands of pieces of porcelain, sponsored the Meissen factory, and intended to fill the Holländische Palais in Dresden–Neustadt with porcelain; this building, which he renamed the Japanische Palais in 1717, might have provided a real-life model for Defoe's porcelain house in *The Farther Adventures of Robinson Crusoe*, the second volume of *Robinson Crusoe*.[47] In light of my analysis of the earlier earthenware episode, the inclusion of a porcelain house in the sequel that Defoe published in 1719 should shed some interesting light on the novel as a whole.

In *The Farther Adventures of Robinson Crusoe*, the aging Crusoe decides to undertake a final adventure abroad. After a sentimental journey to his former island, which is now governed by himself and colonized with human subjects and animals, Crusoe journeys on to the East Indies to explore the northern route to western Europe. After surviving numerous life-threatening situations, he arrives in China in the company of some Jesuit missionaries and merchants. One day, his old Portuguese pilot comes upon a porcelain house and urges him to see this marvelous thing, which he thinks is a "gentleman's house built all with China Ware." Defoe writes:

> Well, said I, are not the Materials of their Building the Product of their own Country; and so it is all China Ware, is not it? No, no, said he, I mean it is a House all made of China Ware, such as you call it in England; or as it is call'd in our Country, Porcellain. Well, said I, such a thing may be; how big is it? Can we carry it in a Box upon a Camel? If we can, we will buy it. Upon a Camel! said the old Pilot, holding up both his Hands, why there is a Family of thirty People lives in it.
>
> I was then curious indeed to see it, and when I came to it, *it was Nothing but this: it was a Timber-House*, or a House built, as we call it in England, with Lath and Plaister, but all the Plaistering was really China Ware, that is to say, *it was plaistered with the Earth that makes China Ware*.
>
> The outside, which the Sun shone hot upon, was glazed, and look'd very well, perfect white, and painted with blue Figures, as the large China Ware in England is painted, and hard as if it had been burnt: As to the inside, all the Walls, instead of Wainscot, were lined up with harden'd and painted Tiles, like the little square Tiles we call Gally-Tiles in England, all made of the finest China, and the Figures exceeding fine indeed, with extraordinary Variety of Colours, mix'd with

Gold, many Tiles making but one Figure, but *join'd so artificially with Mortar,* being made of the same Earth, that it was very hard to see where the Tiles met.[48]

This extraordinary scene of encounter and unmasking may be read as a master allegory of eighteenth-century metaphysics. The goal of Crusoe's unmasking is to show that the chinaware house is not really a chinaware house but a building "plaistered with the Earth that makes China Ware." Since appearances can be deceiving, the building turns out to be nothing more than an ordinary *timber* house, like those built with lath and plaster in England. The glazed surfaces look exquisite on the outside, as do the tiles on the walls of the house inside. But if one were to penetrate the tiled surfaces, one would not expect to find the same kind of materials. Crusoe's penetrating eye does not content itself with the mere appearance of things, so he works to *uncover* the inner truth hidden somewhere else. This metaphysical will to truth implies a necessary act of violence toward the object and reminds us of the trope of "cutting, and opening, and mangling, and piercing" that Swift satirizes so well in "*A Tale of a Tub.*"[49] Inasmuch as Crusoe's act of unmasking turns china into a synecdoche of China (a pun made easy by early modern typography), the violence is directed at both.[50]

Crusoe's unmasking of the lie of the chinaware house and China appeared so successful that a later edition of *Robinson Crusoe* (1815) attaches a footnote, with the hindsight of one hundred years, to gloss the inferior cultural value of porcelain. The footnote contains a lengthy disquisition—with often erroneous technical information—on porcelain making and porcelain's European manufacture and concludes by saying:

> The art of pottery among the Chinese; is one of the most remarkable. But this is a very simple one, and in fact invented by some of the rudest people. They are understood to have an earth possessing certain peculiar virtues in regard to this manufacture; and Barrow informs us, that the merit of their porcelain is less owing to any ingenuity they display in the making of it, than to the prodigious care with which they select the very finest materials, and separate them from all impurities. A very remarkable proof of their want of ingenuity is, that they should have been in possession so long of an art so analogous to that of making glass, and yet should never have been able to invent that beautiful and useful manufacture. Their want of taste in the shapes and ornaments of their vessels, is now proverbial.[51]

This bit of editorial wisdom carries Crusoe's fictional punning on "China Ware" (porcelain and Chinese art) into the unmistakable realm of nineteenth-century *ethnographic realism.* Such realism supports Crusoe's unmasking with the supposedly factual proof provided by Sir John Barrow and other ethnographic travelers of the nineteenth century, contributing directly to the "objective" status of *Robinson Crusoe* as a realist novel. The lie of china in Defoe's text is now clearly spelled out as the lie of China.[52]

But has any such thing as a porcelain house or porcelain room existed in China? The answer is no. Where did Defoe get his idea? Did he make the whole thing up for the sake of exoticizing another place? Not entirely. The probable immediate prototype that comes to mind is the famed chinoiserie Pagoda Tower that Max Emanuel constructed for the park at Nymphenburg on the outskirts of Munich in 1719. This tower was an imitation of the Petit Trianon de porcelaine in Versaille, the earliest chinoiserie garden casino, made at the order of Louis XIV in 1670–1671 and decorated "in the manner of works coming from China."[53] The coincidence of these dates is revealing: *the same year* the Pagoda Tower was constructed, Defoe was writing and publishing his second volume of *Robinson Crusoe*. Also, as I suggested previously, the Holländische Palais in Dresden–Neustadt that Augustus the Strong acquired in 1717 and renamed the Japanische Palais could be another source of inspiration for Defoe's porcelain house.

In 1700, Augustus found himself in desperate straits for means to finance his Swedish war, so he seized the alchemist Böttger, using a military escort to bring him to Dresden, to help transmute lead into gold for him. By 1703 no lead had been transmuted and 40,000 thalers had been wasted on a fraudulent project. Augustus lost patience and placed Böttger under the supervision of Tschirnhaus, whose scientific investigation of the secret of true porcelain was well under way.[54] Tschirnhaus had studied at the University of Leiden from 1668 to 1674 and remained in close contact with the Académie royale des sciences during his early experiments with true porcelain. Much of his work involved melting a variety of highly refractory substances by concentrating radiant heat upon them with a concave iron reflecting mirror.[55] Tschirnhaus also paid several visits to fellow scientists in Italy and France who claimed that they had found the secret of true porcelain. As several scholars have pointed out, this man's preoccupation with ever higher temperatures and his research into the melting point of various refractory substances demonstrate that "he had more than a glimpse of the principles underlying the manufacture of Chinese porcelain."[56]

Using his burning-mirrors, Tschirnhaus experimented with the melting point of asbestos, of a calcium magnesium silicate, and of what appears to have been a kind of kaolin. Apparently, no kilns had been designed in Europe that were capable of reaching the firing temperature of true porcelain at 1,450 degrees celsius. Even though Tschirnhaus did not discover the china clay himself, his early chemical research did move in a direction that James Hutton (1726–1797) would follow several decades later in developing the Plutonist theory of the earth whereby the modern science of geology was founded.[57] When Tschirnhaus was joined by Böttger he had made considerable progress with his mirrors for achieving the maximum heat possible, even though the discovery of kaolin is often attributed to the latter scientist.[58] Together, the two men developed a secret formula that contained alabaster instead of feldspathic rock (thought to be petuntse) as the fusible material to fire with kaolin. The porcelain they manufactured with this formula at the Meissen factory did not, however, exhibit the same degree of whiteness, consistency, and translucency as that of Chinese porcelain. It was not until after 1720—that is, after the death of both men and one year after Defoe published his novel—that alabaster began to be replaced by a feldspathic rock from Sieberlehn (*Sieberlehnstein*) that greatly improved and whitened the body.[59] In that sense, the Meissen factory became the first manufacturer of white porcelain in Europe.

The relationship between glass and white porcelain was a popular subject of speculation in this period, as was amply evidenced by Defoe's novel and the 1815 edition of *Robinson Crusoe* I quoted above, mainly because the renowned French physicist Réaumur took a scientific interest in it. Réaumur was seeking to replicate true porcelain by heating glass packed in refractory powder in saggars in an ordinary potter's kiln. After experimenting with many cements, he came to prefer a mixture of sand and gypsum. He observed that the change in the nature of the glass started at the surface (a fact supporting the alleged chemical change) and grew inward in the form of silklike fibers that were composed of extremely fine grains. The glass could be made entirely granular, like ordinary porcelain, under some heat treatments. Réaumur calls his product "porcelain by transmutation, by revivification, or porcelain from glass." "Glass," said he, "has a polish, a lustre, that is never seen in the fracture of true porcelain. Porcelain is granular, and it is partly by its fine grain that the fracture of porcelain differs from that of terra cotta, and finally it is by the size and arrangement of the grains in them that the kinds of porcelain differ from each other and become closer or less close in nature to glass."[60] The theory behind this experiment is that if metals can be returned to the metallic state after being converted to their calces or dissolved in glass, why should not the sand and stones that gave rise to glass be restorable? As Cyril Stanley Smith has pointed out, the essentially Cartesian corpuscular views that had guided Réaumur's work on steel and iron served equally for porcelain. He was the last scientist for well over a century to have a serious concern with the structure of materials on this level. Réaumur set out to make white porcelain whose appearance and inside would have the same degree of consistency throughout, yet the result was disappointing because, though the inside of his porcelain matched the whiteness of the best Chinese ware, the surface was dark and rough.[61] This seemed to be Réaumur's own unintentional mockery of Crusoe's metaphysical unmasking of porcelain in *The Farther Adventures of Robinson Crusoe*.

As I observed in the beginning of this essay, Defoe's poetics of colonial disavowal was motivated by the economic rivalries of his time. The same can be said of the eighteenth-century scientific research on porcelain, which likewise rode on the wings of global economic forces and was determined to bring about a success story in Europe. Defoe did not live to hear that story. In competition with Chinese export porcelain, Wedgwood and the British ceramic industry, which emerged in the mid-eighteenth century, succeeded in causing the decline of the former on the global market. Fernand Braudel attributes this decline to the natural cycle of fashion in *The Structures of Everyday Life*.[62] He is partially accurate with regard to eighteenth-century chinoiserie and the Franco–German rococo, yet we must not forget that the majority of export porcelain was increasingly intended for daily use either at the dining table, for the toilet, or in the drawing room.[63] The aggressive stance with which Wedgwood marketed its products around the world and even shipped them to China does not seem to substantiate Braudel's argument about fashion and its natural cycle.[64] Moreover, the changing tax laws in Britain during this time also pitted the native industries favorably against the importation of foreign porcelain. As several studies have pointed out, the most important determinant of the favorable balance of trade in Britain during this time was the rise of the percentage of its exports from 8 percent to 20 percent, much of which was due to the existence of Britain's and Europe's colonial economy.[65]

In 1782, Josiah Wedgwood succeeded in developing a pyrometer that could measure degrees of heat above the capacity of existing mercurial thermometers. The invention was said to be the "standard authority in every chemical work for over twenty years, before it was superseded by more accurate instruments."[66] The Scottish geologist James Hall (1761–1832) relied on Wedgwood's pyrometer and porcelain ovens to conduct his laboratory experiments to solve the evidential problem inherent to the field of geological studies. In a scientific paper published in 1806, Hall describes a series of experiments he conducted in a lab setting that proved useful to geology and to chemical science in general. The experiments demonstrated how the most refractory substances could be rendered fusible by repressing the elasticity of what he calls the gaseous parts contained in them. This not only concerned the operations of the mineral kingdom but also the action of fire.[67] To provide credible evidence in support of Hutton's Plutonist theory of the earth—namely, that rocks had been consolidated by heat—Hall's experiments proceeded as if nature's work could be made to replicate itself within a laboratory setting. And that setting was Wedgwood's porcelain kiln.

Wedgwood's own porcelain experiment drew the attention of Erasmus Darwin, a scientist and a renowned poet of his time, who became his close friend. Their friendship evolved into lasting family ties in 1796 when Wedgwood married his daughter Susannah Wedgwood to Robert Waring Darwin.[68] Inspired by Wedgwood's experiments with pottery manufacture, Erasmus Darwin was the first to chant the praises of "the keen-eyed Fire-Nymphs" in his long poem *The Botanic Garden* (1791). Wedgwood, for his part, would fashion novel ceramic designs following (and sometimes revising) the poetic emblems created by the poet Darwin and turning them into sculptured and engraved artifacts. This is duly documented in the family correspondence of Wedgwood, as in the letter dated July 1789 from Josiah Wedgwood to Darwin:

I have now the pleasure of inclosing a cupid, and shall be happy if he meets your approbation. He is drawn and shaded in the stile of the Portland vase . . . nothing can be too good to accompany your charming poem, which like that vase (being both gems of the first water) I admire more and more every time I look upon it.

A thought occurred to me after the Cupid was finished, that, as the Cupid of your botanic garden, he should rather be warming a vegetable than an animal into the passion of love. You will therefore see a detached hand, just sketched with a pencil, holding a flower instead of a moth.[69]

At the beginning of "The Loves of the Plants," part 2 of *The Botanic Garden*, Darwin depicts Cupid holding a flower in his hand, which clearly bears the imprint of his friend's suggestion. For the duration of three decades, such exchanges were frequent and productive and extended to the men's shared interest in chemistry, mineralogy, geology, and art.

In 1788, Wedgwood received samples of Australian clay from the first British colony in New South Wales at the behest of Sir Joseph Banks. He declared that the clay was "an excellent material for pottery and may certainly be made the basis of a valuable manufacture for our infant colony."[70] Wedgwood used this clay to produce a commemorative medallion depicting Hope addressing Peace, Labour, and Plenty. Designed by Henry Webber, this medallion inspired Darwin to eulogize the event in his poem. Wedgwood was one of the first people to receive a copy of this poetic work and wrote to the author: "The speech of Hope is so good and so apropos that I cannot but wish to have it made longer by the same hand, only for *Wedgwood* say *Darwin*, or any other name of 2 syllables, and in that strain go on as long as you please."[71]

As a literary phenomenon, *The Botanic Garden* anticipated what would soon become the dominant "nature" motifs in Romantic poetry before its author's fame was eclipsed by the more enduring impact of Wordsworth, Coleridge, Keats and Shelley.[72] Coleridge hailed Erasmus Darwin as "the first *literary* character of Europe, and the most original-minded man" of his time, but the *topos* of fame and immortality through poetical inscription meets its uncertain fate in *The Botanic Garden* as the work has completely dropped out of the canon of English literature by now.[73] (Not so with Wedgwood ceramics. Even at the close of the millennium, the family torch is still being passed onward by its heir of the eleventh generation, the Honorable Piers Anthony Weymouth, Lord Wedgwood of Barlaston.) If *The Botanic Garden* has lost its appeal as a literary masterpiece, it can be fascinating when read as a work of popular science in verse. The book was probably intended to be read as popular science literature during the heyday of its popularity, because Darwin filled its pages with explanatory notes ranging over all branches of science and technology, and his digressive or allusive notes are often longer than the verse itself. The allusions to the Wedgwood manufactory appear for the most part in canto 2 of part 1, *The Economy of Vegetation*. These allusions not only help unravel the material condition of the eighteenth-century "nature" motifs but invite us to rethink the linkages among science, art, colonial economy, and empire building as, for example, encapsulated in the following:

> GNOMES! as you now dissect with hammers fine
> The granite-rock, the nodul'd flint calcine;
> Grind with strong arm, the circling chertz betwixt,
> Your pure Ka-o-lins and Pe-tun-tses mixt;
> O'er each red saggars burning cave preside,
> The keen-eyed Fire-Nymphs blazing by your side;
> And pleased on WEDGWOOD ray your partial smile,
> A new Etruria decks Britannia's isle.—
> Charm'd by your touch, the kneaded clay refines,
> The biscuit hardens, the enamel shines;
> Each nicer mould a softer feature drinks,
> The bold Cameo speaks, the soft Intaglio thinks. (2.297–310)[74]

The *kaolin* and *petuntse* that are evoked by Darwin's verse may no longer sound the same to our ear as they did to earlier generations, but they belong to a whole class of nature motifs that can be grounded in a simultaneous reading of Wedgwood's ceramic empire, the rise of geological science, and colonial enterprise. This is one of the places where one can begin to comprehend how porcelain achieved its objecthood and identity in the eighteenth century.

In this vein, I'd like to conclude my essay by adding a brief note to the familiar etymology of *china* and the curious definition of this word given by the authoritative dictionary of the English language. According to the *Oxford English Dictionary*, *china* derives from a Persian term, *chini*, that moved through India and eventually made its way into seventeenth-century English. The circulation of this word follows the ancient trading routes and, along with the Portuguese derived etymology of *porcelain*, the etymology of *china* embodies the dual trading histories of England and Portugal in past centuries. The *Oxford English Dictionary* goes on to define *china* as "a species of earthenware of fine semi-transparent texture, originally manufactured in China, and brought to Europe in the 16th c. by the Portuguese, who named it *porcelain*." Apparently, the *Oxford English Dictionary* definition has not benefited from the scientific wisdom of past centuries that would say otherwise. For to define *china* as a species of earthenware is to ignore the history of the scientific experiments in Europe that have thoroughly transformed the meaning and referent of this English word since the sixteenth century. As I have argued in this essay, Crusoe's disavowal of *china* in the earthenware episode and his subsequent unmasking of the chinaware house can only be meaningful insofar as they are seen as part of that history. My reading of *Robinson Crusoe* has attempted to recapture a sense of that history by cross-examining Defoe's life and work through the reverse lens of a Chinese translation that was itself a product of cross-cultural writing. What fascinates me in this process of cross-reading is the revelation of the extraordinary foreignness of Defoe's text. That foreignness is perhaps the site on which the historicity of the familiar is encoded. Like an act of translation, the epistemological encounter with the foreign in the familiar can introduce a radical difference into our reading of an otherwise thoroughly known text and context.

APPENDIX
A GLOSSARY OF CHINESE CHARACTERS

ci	瓷
Da Fu (Defoe)	達孚
Jingdezhen (King-te-Tching)	景德鎮
kaolin	高嶺
Lin Shu	林紓
petuntse	白不子
tao	陶
tseki	瓷器
wa	瓦
waqi	瓦器
Yin Hongxu (d'Entrecolles)	殷弘緒
Yixing	宜興
Yuanming yuan	圓明園
Zeng Zonggong	曾宗鞏

Notes

The writing of this chapter benefited enormously from the input of the editors of *Critical Inquiry* in which an earlier version was published. I thank the following individuals for their generous comments on my work at different stages: Elizabeth Helsinger, Michael Fischer, Roger Hart, Paul Rabinow, Judith Klein, Mary Campbell, Marta Hanson, Chandra Mukerji, Neil De Marchi, Lee Ou-fan Lee, and Tuo Li. I am also grateful to Noah Heringman, Larissa Heinrich, and Jami Proctor-Xu for their assistance in the research process as well as the preparation of this manuscript.

1. "Robinson Crusoe," in *The Second Common Reader* (New York: Harcourt, Brace, and World, 1960), 48–49.

2. The classical political economists from the eighteenth century onward often used Robinson Crusoe as their favorite model of illustration. The solitary individual on a desert island served as a convenient starting point for building their systems. Concerning the "primitive" character of Crusoe's production, Marx offers a sarcastic comment:

> Necessity itself compels him to apportion his time accurately between different kinds of work. Whether one kind occupies a greater space in his general activity than another, depends on the difficulties . . . to be overcome in attaining the useful effect aimed at. This our friend Robinson soon learns by experience, and having rescued a watch, ledger, and pen and ink from the wreck, commences, like a true-born Briton, to keep a set of books. (*Capital: A Critique of Political Economy*, trans. Samuel Moore and Edward Aveling, 3 vols. [New York: International Publishers, 1967], 1:76–77)

Marx criticizes David Ricardo for using Crusoe in this manner:

> He makes the primitive hunter and the primitive fisher straightway, as owners of commodities, exchange fish and game in the proportion in which labour-time is incorporated in these exchange-values. On this occasion he commits the anachronism of making these men apply to the calculation, so far as their implements have to be taken into account, the annuity tables in current use on the London Exchange in the year 1817. "The parallelograms of Mr. Owen" appear to be the only form of society, besides the bourgeois form, with which he was acquainted (1:81 n).

Insisting on the social nature of economic production, Marx rejected Adam Smith's and Ricardo's fiction of *homo economicus*, pointing out that

> the individual and isolated hunter or fisher who forms the starting point with Smith and Ricardo, belongs to the insipid illusions of the eighteenth century. They are Robinsonades which do not by any means represent, as students of the history of civilization imagine, a reaction against over-refinement and a return to a misunderstood natural life. They are no more based on such a naturalism than is Rousseau's "contrat social," which makes naturally independent individuals come in contact and have mutual

intercourse by contract. They are the fiction and only the aesthetic fiction of the small and great Robinsonades. (*A Contribution to the Critique of Political Economy,* trans. N. I. Stone [Chicago: C. H. Kerr, 1913], 265–66)

3. *ReOrient: Global Economy in the Asian Age* (Berkeley and Los Angeles: U of California P, 1998), 282.

4. See J. A. Lloyd Hyde and Ricardo R. Espirito Santo Silva, *Chinese Porcelain for the European Market* (1956; Lisbon: Fundacão Ricardo do Espirito Santo Silva, 1994); John Goldsmith Phillips, *China Trade Porcelain: An Account of Its Historical Background, Manufacture, and Decoration and a Study of the Helena Woolworth McCann Collection* (Cambridge, MA: Harvard UP, 1956); and David Howard and John Ayers, *China for the West: Chinese Porcelain and Other Decorative Arts for Export Illustrated from the Mottahedeh Collection,* 2 vols. (London: Sotheby Parke Bernet, 1978).

5. Defoe was reacting to the birth of consumerism in the early eighteenth century, which also saw the rise of European chinoiserie. Werner Sombart and Fernand Braudel diagnosed chinoiserie as the conspicuous consumption of luxury in the early stages of capitalism. See Werner Sombart, *Luxury and Capitalism,* trans. W. R. Dittmar (1913; Ann Arbor: U of Michigan P, 1967); and Fernand Braudel, *The Structures of Everyday Life,* trans. Siân Reynolds (Berkeley and Los Angeles: U of California P, 1992). Bruce P. Lenman's recent study shows that the English population on both sides of the Atlantic experienced "its first great wave of consumerism, in which imported Asiatic products and manufactures played an important role" ("The English and Dutch East India Companies and the Birth of Consumerism in the Augustan World," *Eighteenth-Century Life* 14 [Febuary 1990]: 62).

6. *Daniel Defoe: His Life* (Baltimore: Johns Hopkins UP, 1989), 64–65.

7. *A Tour through the Whole Island of Great Britain (1724–1726)* (London: Dutton, 1962), 1:165–66.

8. Indeed, chinaware became a trope that could figure other kinds of difference as well. For example, John Gay, Defoe's contemporary, wrote a satirical poem entitled "To a Lady on her Passion for Old China" in 1725. In it, womanhood and porcelain evoke each other metonymically and synecdochically, whereas manhood is equated to earthenware— rough on the surface but sturdy on the inside. Gay's poem spells out an aesthetics of materiality categorically grounded in the metaphysics of appearance and reality, surface and depth, femininity, masculinity, and so on.

9. Peter Hulme, *Colonial Encounters: Europe and the Native Caribbean, 1492–1797* (1986; London: Routledge, 1992), 176; hereafter abbreviated *CE.*

10. Ian Watt, *The Rise of the Novel: Studies in Defoe, Richardson, and Fielding* (Berkeley and Los Angeles: U of California P, 1957), 65; quoted in *CE,* 176.

11. By Hulme's account, the two main studies usually invoked for this pervasive spiritual motif are G.A. Starr, *Defoe and Spiritual Autobiography* (Princeton: Princeton UP, 1965); and J. P. Hunter, *The Reluctant Pilgrim: Defoe's Emblematic Method and the Quest for Form in Robinson Crusoe* (Baltimore: Johns Hopkins UP, 1966).

12. "Daniel Defoe," ed. and trans. Joseph Prescott, *Buffalo Studies* 1 (1964): 24–25; quoted in *CE,* 216.

13. *Émile ou de l'Éducation,* in *Oeuvres complètes de J. J. Rousseau,* 13 vols. (Paris: Hachette, 1887–1912), 2:156, emphasis added, hereafter abbreviated *EO;* trans. Barbara Foxley, *Émile* (1911; London: J. M. Dent, 1963), 147, emphasis added; hereafter abbreviated *E.*

14. Quoted in Peter Costello, *Jules Verne: Inventor of Science Fiction* (New York: Scribner, 1978), 94. Peter Redfield drew my attention to the Verne legacy and I thank him here.

15. I thank Caren Kaplan and Eric Smoodin for mentioning these films to me.

16. *A Theory of Literary Production,* trans. Geoffrey Wall (London: Kegan Paul, 1978), 241–42. Incidentally, the quote might be useful in helping us think about J. M. Coetzee's novel *Foe* (New York: Penguin, 1987).

17. Ibid., 233.

18. Shortly before d'Entrecolles, Le Comte had also written about Chinese porcelain and its manufacture in 1697.

19. See below for further discussion.

20. Defoe's pantiles were S-shaped, varnished, and then glazed and had a fine red color with good texture, which suggests a Dutch import. See Backscheider, *Daniel Defoe,* 64.

21. Daniel Defoe, *Robinson Crusoe,* ed. Michael Shinagel, 2nd ed. (New York: Norton, 1994), 88; emphasis added.

22. It is interesting that Crusoe mentions sand because sand was among the several materials, such as chalk, gypsum, alabaster, and ferruginous stone, upon which eighteenth-century scientists experimented to approximate the translucent texture of true porcelain.

23. Lin Shu rendered and published the first two volumes of *Robinson Crusoe* in 1905–1906. This happened at a time when the various English editions of *Robinson Crusoe* being published were omitting the second and third volumes. Volume 2 is *The Farther Adventures of Robinson Crusoe, Being the Second and Last Part of His Life, and of the Strange, Surprizing Accounts of His Travels Round Three Parts of the Globe* in which Crusoe makes a trip to the East Indies and recounts his adventures in South Asia, China, and Russia. See below for my discussion of Crusoe's alleged encounter with a porcelain house in China.

24. From a personal point of view, this has been one of the most rewarding translingual encounters I have come across. For a discussion of the epistemological implications of the recent encounter between Chinese, English, and other languages, see my book *Translingual Practice: Literature, National Culture, and Translated Modernity* (Stanford: Stanford UP, 1995), xv–42.

25. *Robinson Crusoe,* 89.

26. Although the distinction between *wa,* on the one hand, and *tao* and *ci* on the other is clear, Lin Shu uses the words *ci* and *tao* interchangeably, as was often the case in early Chinese documents on porcelain. See Jiang Qi, *Taoji* (Records of pottery and porcelain) (ca. 1322–1325); Zhu Yan, *Taoshuo* (Description of pottery and porcelain) (1774); and Lan Pu, *Jingdezhen taolu* (History of pottery and porcelain in Jingdezhen) (1815). Modern Chinese scholars tend to make a sharper distinction between *tao* and *ci.* Fu Zhenlun, for example, renders Jiang Qi's *tao* (porcelain of Jingdezhen) as *ci* in his translation of Jiang's classical text into modern Chinese. See his *Zhongguo gu taoci luncong* (Essays in ancient Chinese pottery and porcelain), (Beijing, 1994), 178. See also Fu's *Mingdai ciqi gongyi* (The art of porcelain making in the Ming dynasty) (Beijing, 1955); and Xiong Liao, *Zhongguo taoci meishu shi* (History of the art of pottery and porcelain in China), (Beijing, 1993).

27. "The Task of the Translator," *Illuminations*, ed. Hannah Arendt, trans. Harry Zohn (New York: Schocken Books, 1969), 78.

28. See Benjamin, "Die Aufgabe des Übersetzers," *Illuminationen: Ausgewählte Schriften*, ed. Siegfried Unseld (Frankfurt a. M.: Suhrkamp, 1961), 65. My speculation about *Gefäß* is also supported by Benjamin's use of the image of a clay bowl (*Tonschale*) in a similar manner in his essay "Der Erzähler: Betrachtungen zum Werk Nikolai Lesskows." In it, Benjamin observes:

> Die Erzählung, wie sie im Kreis des Handwerks—des bäuerlichen, des maritimen und dann des städtischen—lange gedieh, ist selbst eine gleichsam handwerkliche Form der Mitteilung. Sie legt es nicht darauf an, das pure "an sich" der Sache zu überliefern wie eine Information oder Rapport. Sie senkt die Sache in das Leben des Berichtenden ein, um sie wieder aus ihm hervorzuholen. So haftet an der Erzählung die Spur des Erzählenden wie die Spur der Töpferhand an der Tonschale. [The storytelling that thrives for a long time in the milieu of the work—the rural, the maritime, and the urban—is itself an artisan form of communication, as it were. It does not aim to convey the pure essence of the thing, like information or a report. It sinks the thing into the life of the storyteller, in order to bring it out of him again. Thus traces of the storyteller cling to the story the way the handprints of the potter cling to the clay vessel."] ("Der Erzähler: Betrachtungen zum Werk Nikolai Lesskows," *Illuminationen*, 418; "The Storyteller: Reflections on the Works of Nikolai Leskov," *Illuminations*, 91–92)

29. It is interesting that Jacques Derrida seizes on this image in his own elaboration on Benjamin's theory of translation. See "Des Tours de Babel," ed. and trans. Joseph F. Graham, in *Difference in Translation* (Ithaca: Cornell UP, 1985), 165–207.

30. "Die Aufgabe des Übersetzers," 80.

31. D'Entrecolles, letter to Père Orry, September 1, 1712, *Lettres édifantes et curieuses, écrites des missions éstrangères,* ed. Charles Le Gobien et al., 28 vols. in 26 (Paris: N. Le Clerc, 1707–1758), 12:272.

32. Crusoe's repeated use of the word *earthenware* appeared confusing to Lin and his English-language informant Zeng. They managed to replace most instances of this word with *ci* and *tao* except in one case in which they could not decide whether Crusoe makes porcelain or a mixture of porcelain *and* inferior pottery. Hence, the word *wayu* (sunbaked vessel) is used once along with *ci*. The word *wa* is used both as an adjective and as a noun, with a difference in meaning. As an adjective, it means "earthenware," as in *wayu* (earthenware vessel) and *wapen* (earthenware pot); as a noun, it means "tile," as in *liuli wa* (glass tile), *ciwa* (porcelain tile), and so forth.

33. Da Fu (Defoe), *Lu Bingsun piaoliuji* (Robinson's Adventures), trans. Lin Shu and Zeng Zonggong, 2 vols. (Shanghai, 1905), 1:106. The word *ciwa* is glossed as "porcelain tile," but it could also reflect Lin and Zeng's indecision over *ci* and *wa* as used in Defoe's text.

34. The rise of the rococo style was greatly indebted to chinoiserie and was grasped by contemporaries as distinctly modern. It arose because of the expansion of global trade and the emergence of a new market in Europe whereby the artist–artisan producers of

rococo work (cabinetmakers, jewelers, engravers) were to some extent producing for themselves and each other. Patricia Crown points out that "'modern' was an exact synonym for Rococo" ("British Rococo as Social and Political Style," *Eighteenth-Century Studies* 23 [Spring 1990]: 274).

35. Yixing ware was named after the town Yixing in the province of Jiangsu where such red-tinted ware was produced. This ware became fashionable in China in the sixteenth and seventeenth centuries and was widely collected by the literati. See Fu, *Zhongguo gu taoci luncong* and *Mingdai ciqi gongyi*. For a technical analysis of traditional Yixing ware in English, see Sun Jing, Ruan Meiling, and Gu Zujun, "Microstructure of Ancient Yixing Zisha Ware Excavated from Yangjiao Hill," *Scientific and Technological Insights on Ancient Chinese Pottery and Porcelain: Proceedings of the International Conference on Ancient Chinese Pottery and Porcelain Held in Shanghai from November 1 to 5, 1982*, ed. Shanghai Institute of Ceramics, Academia Sinica (Beijing: Science Press, 1986), 86–90.

36. In the literature of the period, as in Gay's poem mentioned in note 8, the difference between chinaware (porcelain) and earthenware was constantly exploited as a metaphor.

37. For a study of English earthenware and its Dutch and Chinese connections, see A. H. Church, *English Earthenware: A Handbook to the Wares Made in England during the Seventeenth and Eighteenth Centuries as Illustrated by Specimens in the National Collections* (1884; London: Wyman and Sons, 1904).

38. *From London to Land's End* (London: Cassell, 1892), 18; emphasis added.

39. *Eighteenth-Century German Porcelain* (London: Rockliff, 1958), 19.

40. In the Tang dynasty, porcelain was described as "fake/simulated jade" (Fu, *Zhongguo gu taoci luncong*, 182).

41. Quoted in Michel Beurdeley, *Chinese Trade Porcelain*, trans. Diana Imber (Rutland, VT: C. E. Tuttle, 1962), 10.

42. "*The Rape of the Lock:* An Unnoticed Significance of 'China's Earth,'" *Notes and Queries*, ns 34 (September 1987): 342.

43. See Church, *English Earthenware*, 44–55. For detailed information on the early experiments of Dwight and the Elers brothers, see Eliza Meteyard, *The Life of Josiah Wedgwood: From His Private Correspondence and Family Papers in the Possession of Joseph Mayer, Esq., F.S.A., F. Wedgwood, Esq., C. Darwin, Esq., M.A., F.R.S., Miss Wedgwood and Other Original Sources*, 2 vols. (London: Hurst and Blackett, 1865–1866).

44. For a detailed archival study of the historical interactions among Dutch and Chinese traders and manufacturers, see T. Volker, *Porcelain and the Dutch East India Company: As Recorded in the "Dagh-Registers" of Batavia Castle, Those of Hirado and Deshima and Other Contemporary Papers, 1602–1682* (Leiden: E. J. Brill, 1954). For related studies, see Carolyn Saville Woodward, *Oriental Ceramics at the Cape of Good Hope, 1662–1795: An Account of the Porcelain Trade of the Dutch East India Company with Particular Reference to Ceramics with the V.O.C. Monogram, the Cape Market, and South African Collections* (Cape Town: Balkama, 1974).

45. Backscheider, *Daniel Defoe*, 64, 65.

46. See Jan Divis, *European Porcelain*, trans. Iris Urwin (New York: Excalibur, 1983), 26–27; and Savage, *Eighteenth-Century German Porcelain*, 28, 52–56.

47. Volume 2, *The Farther Adventures of Robinson Crusoe*, was published by William Taylor on August 20, 1719, and volume 3, *Serious Reflections during the Life and Surprising*

Adventures of Robinson Crusoe: With His Vision of the Angelick World, appeared on August 6, 1720.

48. *The Farther Adventures of Robinson Crusoe, Being the Second and Last Part of His Life, and of the Strange, Surprizing Accounts of His Travels Round Three Parts of the Globe,* vol. 2 of *The Life and Strange Surprizing Adventures of Robinson Crusoe, of York, Mariner: Who Lived Eight and Twenty Years, All Alone in an Un-Inhabited Island on the Coast of America, Near the Mouth of the Great River Oroonoque, Having Been Cast on Shore by Shipwreck, Wherein All the Men Perished but Himself. With an Account How He Was at Last As Strangely Deliver'd by Pyrates* (London: Printed for W. Taylor, 1719), 310–11; emphasis added.

49. *"A Tale of a Tub," to Which is Added "The Battle of the Books" and "The Mechanical Operation of the Spirit,"* 2nd ed. (Oxford: Clarendon, 1958), 173.

50. Commenting on China's Great Wall in volume 2, Crusoe observes: "Do you think it would stand out an army of our country people, with a good train of artillery; or our engineers, with two companies of miners? Would they not batter it down in ten days, that an army might enter in battalia, or blow it up into the air, foundation and all, that there should be no sign of it left?"(182). Defoe was an uncanny prophet when he put these words into the mouth of his character. The Anglo–Chinese wars in the nineteenth century would bear him out in Britain's premeditated destruction of the very physical symbols, if not the Great Wall itself, that the British had believed were the pride of Chinese civilization: the Yuanming Yuan imperial gardens, temples, and palaces.

51. Defoe, *Robinson Crusoe,* the *Naval Chronicle* ed. (London: J. Mawman, 1815), 433–34.

52. The evocation of glass is interesting because it runs counter to what eighteenth-century scientists, such as the French physicist Réaumur, would say about the relatively inferior technology of glassmaking.

53. Phillips, *China-Trade Porcelain,* 47.

54. Detailed information is found in Savage's discussion of the birth of the Meissen factory in *Eighteenth-Century German Porcelain,* 22–92.

55. Refractory substances result from chemical changes in a mixture of feldspar, granite, and pegmatite. They are an essential element of true porcelain, called *kaolin* in China after the name of the hills near Jingdezhen. This material fires white and composes the body of the paste. The essential fusible element is supplied by *petuntse,* which fuses under heat into a kind of natural glass.

56. Ibid., 24.

57. See below for a discussion of porcelain-related chemical experiments conducted by Réaumur. For a recent study of Hutton and his theory of the earth, see Dennis R. Dean, *James Hutton and the History of Geology* (Ithaca: Cornell UP, 1992).

58. See Savage, *Eighteenth-Century German Porcelain,* 25.

59. See ibid., 36.

60. Quoted in Cyril Stanley Smith, "Porcelain and Plutonism," *Toward a History of Geology,* ed. Cecil J. Schneer (Cambridge: MIT Press, 1969), 324, 323.

61. See ibid., 326.

62. *The Structures of Everday Life,* 186.

63. For a detailed study of inventories and markets for such items beginning in the 1720s, see Beurdeley, *Porcelain of the East India Companies*, 39–47.

64. For a recent study of the rise of Wedgwood, see Neil McKendrick, "Josiah Wedgwood and the Commercialization of the Potteries," *The Birth of a Consumer Society: The Commercialization of Eighteenth-Century England*, ed. Neil McKendrick, John Brewer, and J. H. Plumb (London: Hutchinson, 1983), 100–45.

65. See Backscheider, *Daniel Defoe*, 532. See also, W. A. Speck, *Stability and Strife: England, 1714–1760* (Cambridge, MA: Harvard UP, 1979), 126; George Rudé, *Hanoverian London, 1714–1808* (Berkeley and Los Angeles: U of California P, 1971), 20–24, 32–35; and T. S. Ashton, *An Economic History of England: The Eighteenth Century* (London: Methuen, 1955), 151.

66. Ann Finer and George Savage, eds., *The Selected Letters of Josiah Wedgwood* (London: Cory, Adams, and Mackay, 1965), 264.

67. See James Hall, "Account of a Series of Experiments, Showing the Effects of Compression in Modifying the Action of Heat," *Transactions of the Royal Society of Edinburgh*, vol. 6 (Edinburgh: The Society, 1806), 71–186.

68. Charles Darwin, who also married a Wedgwood daughter, was Erasmus Darwin's grandson. See Finer and Savage, *Selected Letters*, 342. Also see Eliza Meteyard, *The Life of Josiah Wedgwood*, 2 vols. (London: Hurst and Blackett, 1866), 2:559.

69. Josiah Wedgwood to Erasmus Darwin, July 1789 (Etruria Collection. 19003–26. The Library of the University of Keele, Staffordshire).

70. Josiah Wedgwood to Sir Joseph Banks, March 13, 1790 (Royal Society MSS. Lp.9.167).

71. Josiah Wedgwood to Erasmus Darwin, June 28, 1789 (Etruria Collection. 19001–26).

72. For a new study of Romanticism from this perspective, see Noah Heringman, *Romantic Rocks, Aesthetic Geology* (Ithaca: Cornell UP, forthcoming 2004).

73. See Desmond King-Hele's editorial note to Erasmus Darwin, *The Botanic Garden* (1791; Menston, UK: Scolar Press, 1973), i. Cf. the discussion of Darwin in Rachel Crawford's essay (chapter 7) in the present volume.

74. Erasmus Darwin, *The Botanic Garden* (London: Joseph Johnson, 1791), 57–58. Darwin goes on here to eulogize the pathos of the Wedgwood cameo depicting "the poor fetter'd Slave, on bended knee,/From Britain's sons imploring to be free" (2:315–16).

CHAPTER SIX

Frankenstein, Racial Science, and the "Yellow Peril"

ANNE K. MELLOR

On that most famous of fictional nights, that "dreary night of November" when Victor Frankenstein beheld the accomplishment of his scientific efforts to create a human being, to bestow animation upon lifeless matter and thus reveal the origin of life itself, he looked upon the results of his experiment with undiluted horror.

> [B]y the glimmer of the half-extinguished light, I saw the dull yellow eye of the creature open; it breathed hard, and a convulsive motion agitated its limbs.
>
> How can I describe my emotions at this catastrophe, or how delineate the wretch whom with such infinite pains and care I had endeavoured to form? His limbs were in proportion, and I had selected his features as beautiful. Beautiful!— Great God! His yellow skin scarcely covered the work of muscles and arteries beneath; his hair was of a lustrous black, and flowing; his teeth of a pearly whiteness; but these luxuriances only formed a more horrid contrast with his watery eyes, that seemed almost of the same colour as the dun white sockets in which they were set, his shrivelled complexion, [and] straight black lips.[1]

Many readers have rightly attributed Victor's horror to the fact that the Creature has now assumed a life, an agency of its own, an independence that threatens and makes demands on Victor's own life. Others have called attention to the ugliness of the Creature and its gigantic size. Still others have seen the Creature as one of the undead, the abjected, a liminal figure who transgresses and thus denies the boundaries between life and death. Victor Frankenstein himself identifies the horror of the creature's countenance with an embalmed Egyptian mummy: "A mummy again endued with animation could not be so hideous as that wretch" (*F* 53). And the ship captain Walton, when he finally sees the Creature at close range, also thinks immediately of a mummy: "As he hung over the coffin [of Victor Frankenstein], his face was concealed by long locks of ragged hair; but one vast hand was extended, in colour and apparent texture, like that of a mummy" (216).

The terror aroused throughout the novel by the Creature's appearance, not only in Frankenstein and Walton, but in all who see him—from passing villagers to the De Lacey family—has yet another source, I believe, and one whose full significance has been overlooked. Recall precisely how Mary Shelley describes this newborn giant: his skin is *yellow,* his hair is "black and flowing," both the irises of his eyes and the sockets in which they are set are "dun white" or light gray-brown. This Creature is not white skinned, not blond haired, not blue eyed. He is not Caucasian. He is not of the same race as his maker, Victor Frankenstein, who, as opposed to the Creature, "lies *white* and cold in death" (219, emphasis added).[2]

The observation that the Creature is of a different skin color and hence of a different race is made for the reader long before we enter Frankenstein's attic laboratory and witness the climax of Frankenstein's experiment. The first glimpse we as readers have of the Creature is in the opening pages of the novel when Walton and his men, seeking a passage to China through the North Pole, their ship trapped in an ice floe, see at a distance of a half mile "a low carriage, fixed on a sledge and drawn by dogs, pass on towards the north," carrying "a being which had the shape of a man, but apparently of gigantic stature" (17). The very next morning, they rescue the stranded Victor Frankenstein, whom Walton describes as "not, as the other traveller seemed to be, a savage inhabitant of some undiscovered island, but an European" (18). The Creature then, is not an European, not Caucasian, but of some other race, like those found by Captain James Cook in his famous voyages to the Pacific Islands or by the East India Company's trading ships sailing among the spice islands in the Indian Ocean. But this "savage" inhabits an island *north* of the "wilds of Tartary and Russia" whence Frankenstein has pursued him, north of Archangel, the northernmost city in western Asia from which Walton has set sail (200).

A yellow-skinned man crossing the steppes of Russia and Tartary, with long black hair and dun-colored eyes—most of Shelley's nineteenth-century readers would immediately have recognized the Creature as a member of the Mongolian race, one of the five races of man first classified in 1795 by Johann Friedrich Blumenbach, the scholar who, more than any other, founded the modern science of physical anthropology. In constructing Frankenstein's creature as a member of a distinctly different, non-Caucasian race, Shelley again located her novel at the cutting edge of early-nineteenth-century scientific research. I have elsewhere discussed the ways in which she drew on other contemporary scientific discoveries—Luigi Galvani's theory of animal electricity, Humphry Davy's discoveries in chemistry, and Erasmus Darwin's theory of animal and human evolution through sexual selection—in order to define Frankenstein's project as interventionist or "bad" science and his creature as an antievolutionary species composed of both human and animal organs.[3] Here I want to explore the ways in which she used insights garnered from the new scientific field of ethnology or natural history to suggest yet another way in which Victor has contradicted his own goal, the creation of a "new" and better species, in his construction of the Creature.

Let me briefly summarize eighteenth-century concepts of the origin and character of the human species. By 1800, two opposing theories explaining the major differences between human "tribes" or "nations" were in wide circulation in Europe. One theory, the polygenist theory, reaffirmed by such thinkers as Immanuel Kant, Lord Monboddo, and Charles White, held that human tribes had originated independently one from another and could be placed in clearly demarcated degrees along the great Chain of Being. The "missing link" between human beings and apes was filled, according to these writers, by black Africans, who constituted a distinct and different species. In opposition to this biological determinism, many scholars followed Christian doctrine and the writings of Aristotle, Alexander, and Strabo in arguing that all the human "tribes" or nations were members of the same human species or Family of Man, descended through time from one original couple (Adam and Eve). Differences in skin and hair color, skull formation, and anatomy were attributed to organic alteration caused in large part by differing *environmental* conditions, such as climate, water, air, and food. The most famous and widely respected proponent of this monogenist view was of course Comte de Buffon, who both in his *Variétés dans l'espèce humaine* (1749) and again in his forty-four-volume *Histoire naturelle* (1749–1804) argued that animals who can procreate together, and whose progeny can procreate, are of one species. Buffon therefore concluded that all humans belong to the same species but that "climate, food, manners and customs produce not only a difference in sentiment, but even in the external form of different people."[4]

It was Johann Friedrich Blumenbach who in his 1775 doctoral dissertation, entitled *De generis humani nativa varietate*, developed the analytical concept of race to classify the specific varieties or subgroups within the human species. As both Nicholas Hudson and Ivan Hannaford have demonstrated, the concept of race as a stable, transnational, biological or genetic category was not widely available before the late eighteenth century.[5] In 1775, Blumenbach—following Renaissance and Linnaean theories of the four humors—identified *four* races of man: Caucasian (a term he coined, drawing on recent theories that Indo–European languages originated in the region of the Caucasus), Mongolian, American, and Ethiopian; in the second edition (1781), he added a fifth race, the Malay. And in the third edition (1795), Blumenbach not only greatly expanded his descriptions of these five races of man but also located them in historical order.[6] The Caucasian race was the oldest or most primitive race. The Mongolian and Ethiopian races were at the opposite extreme, the two races furthest removed in historical time or "degenerated" from the originary Caucasian race (for Blumenbach, "de-genera-tion" is a purely technical term meaning "generated later in time"). In between, as historically "transitional" races, stood the American—halfway between the Caucasian and the Mongolian—and the Malay—halfway between the Caucasian and the Ethiopian.

Blumenbach based his racial classifications on four characteristics: skin color, hair, the shape of the skull, and physical anatomy or body form. In 1775 he had followed Linnaeus in classifying four racial varieties—Caucasian, Asian, American and African—asserting that "there is an almost insensible and indefinable transi-tion from the pure white skin of the German lady through the yellow, the red, and the dark nations, to the Ethiopian of the very deepest black" (*Mankind* 107). Ex-panding to five distinct races in 1795, in a section entitled "On the Causes and Ways by which Mankind has Degenerated, as a Species," Blumenbach argued that skin color was the most reliable indicator of racial identity. He then lists the five racial skin colors as follows:

1. The white color holds the first place, such as is that of most European peoples. . . .
2. The second is the *yellow, olive-tinge*, a sort of colour half-way between grains of wheat and cooked oranges, or the dry and exsiccated rind of lemons: very usual in the Mon-golian nations.
3. The *copper colour* or dark orange, or a sort of iron, not unlike the bruised bark of cinna-mon or tanner's bark: peculiar to the Americans.
4. *Tawny*, mid-way between the colour of fresh mahogany and dried pinks or chestnuts: common to the Malay race and the men of the southern Archipelago.
5. Lastly, the *tawny-black*, up to almost a pitchy blackness (*jet-black*), principally seen in some Ethiopian nations. (*Mankind* 209)

To this list Blumenbach then added the differences in hair, as follows:

1. The first of a brownish or nutty colour, shading off on the one side into yellow, on the other into black; soft, long, and undulating. Common in the nations of temperate Europe; . . .
2. The second, black, stiff, straight, and scanty; such as is common to the Mongolian and American nations.
3. The third, black, soft, in locks, thick and exuberant; such as the inhabitants of most of the islands of the Pacific Ocean exhibit.
4. The fourth, black and curly, which is generally compared to the wool of a sheep; com-mon to the Ethiopians. (224)

Blumenbach continued his lists of racial differences, invoking Petrus Camper's 1791 theory of the seven gradations in facial angle (see Figure 6.1, a chart taken from Camper's English follower, Charles White), recent phrenological analyses of skull formation, and current studies of anatomical differences—an area in which he was already a renowned specialist.[7]

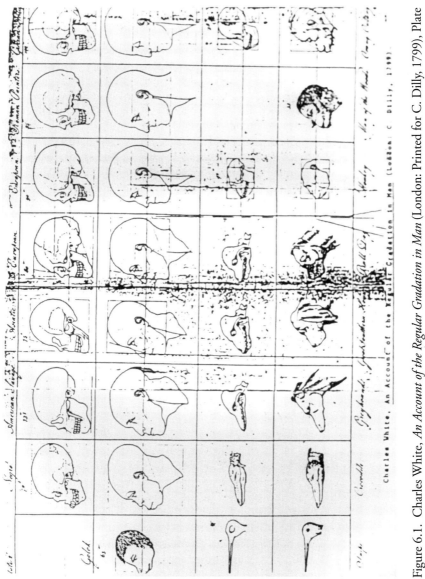

Figure 6.1. Charles White, *An Account of the Regular Gradation in Man* (London: Printed for C. Dilly, 1799), Plate II. Engraving.

Blumenbach's racial classifications were immediately disseminated in England. Lord Kames endorsed Blumenbach's argument that there were "different races or kinds of men . . . naturally fitted for different climates" in the third "considerably improved" and widely read edition of his *Sketches of the History of Man*, which inspired the three younger royal princes to attend Blumenbach's lectures in Göttingen during the winter of 1786.[8] Blumenbach's theory of the five races of mankind was then incorporated by Kames's American disciple, Samuel Stanhope Smith, into the second "enlarged and improved" edition of his *Essay on the Causes of the Variety of Complexion and Figure in the Human Species* published in New York and London in 1810.[9]

Blumenbach's two most ardent English disciples were James Cowles Prichard, a prominent Welsh physician at the Medical Infirmary in Bristol, and William Lawrence, professor of anatomy and surgery at St. Bartholomew's Hospital in London. Prichard dedicated the second and third editions of his widely read *Researches into the Physical History of Man*, first published in 1813, to Blumenbach, even though he followed Dr. James Hunter in identifying not five but seven "varieties" of man. Included among them were the "yellow and bald-headed Mongoles" as the fourth variety whose complexion, he added, is "of a yellowish tint passing into an olive, and stiff long black hair."[10] Prichard, a devout Evangelical Christian like Blumenbach, was a fierce proponent of monogenism, the belief that "all mankind constitute but one race [here meaning species] or proceed from a single family' (i.e., Adam and Eve).[11]

It was William Lawrence who most vigorously promoted Blumenbach's materialist belief in the existence of five distinctively different human races, first in his annual courses of lectures in anatomy at St. Bartholomew's Hospital, which began in 1812 and drew heavily on Blumenbach's *Comparative Anatomy* (1805), a treatise that Lawrence had translated and expanded in 1807 during his early professional career as a demonstrator in anatomy under Dr. John Abernethy. Lawrence then expanded and published these lectures in 1819 in a volume entitled *Lectures on Physiology, Zoology, and the Natural History of Man*, which he dedicated to Blumenbach.

I focus on Lawrence because, as we shall see, he was a close friend of both Percy and Mary Shelley. Lawrence was a strong advocate of the mechanistic concept that the living human organism was entirely composed of more or less dense bodily matter (as opposed to Abernethy's vitalist belief in an independent life or "soul" that escaped the body at death).[12] Lawrence—who served as one of the models for Professor Waldman in Mary Shelley's *Frankenstein*—believed that chemistry is the ultimate science, one that "teaches us the composition of bodies."[13] In his *Lectures*, Lawrence followed Blumenbach closely in insisting that the human species was distinctly different from all animal species, including the apes; that the African or Ethiopian race was not a separate species or missing link on a chain of being between apes and humans but rather a variety of the human species; and that there were indeed five races of man, based on skin color, hair type, eye color, skull formation, and anatomy.

But Lawrence made one all-important addition to Blumenbach's racial theory. Following Kant and Herder in Germany, Lawrence attributed specific *moral characteristics* to each racial type. I cannot give you a complete list of all the intellectual and emotional qualities that Lawrence, drawing on Buffon, Linnaeus, and many others, identified with each race.[14] In his view the white race has preeminence "in moral feelings and mental endowments"—although each lesser race has positive as well as negative attributes (*Lectures* 476). Here I quote his description only of the yellow-skinned Mongolian race:

> The Mongolian people differ very much in their docility and moral character [from the savage tribes of North America]. While the empires of China and Japan prove that this race is susceptible of civilisation, and of great advancement in the useful and even elegant arts of life, and exhibit the singular phenomenon of political and social institutions between two and three thousand years older than the Christian era, the fact of their having continued nearly stationary for so many centuries, marks an inferiority of nature and a limited capacity in comparison to that of the white races.
>
> When the Mongolian tribes of central Asia have been united under one leader, war and desolation have been the objects of the association. Unrelenting slaughter, without distinction of condition, age, or sex, and universal destruction have marked the progress of their conquests, unattended with any changes or institutions capable of benefiting the human race, unmingled with any acts of generosity, any kindness to the vanquished, or the slightest symptoms of regard to the rights and liberties of mankind. The progress of ATTILA, ZINGHIS [Genghis Khan], and TAMERLANE, like the deluge, the tornado and the hurricane, involved every thing in one sweeping ruin. (*Lectures* 483)

Here Lawrence attempted to provide a scientific basis for what Edward Said has since called "orientalism," the cultural production of a racial stereotype of the Mongol or Asian race as on the one hand, "stationary" or culturally stagnant, lazy, mired in luxury and decadence, and on the other hand, innately violent, barbaric, and destructive.[15] Lawrence, of course, was recalling the thirteenth-century conquest of Asia, from China west to Russia, Poland, and Hungary, by Genghis Khan, his sons and grandsons, and the army known in European annals as "the Golden Horde."[16]

Lawrence made two further points of possible relevance to Shelley's representation of Frankenstein's creature. First, given his belief in the preeminence of the Caucasian race and the distinct inferiority of the black or Ethiopian race, when he took up the fact that the ancient Egyptians had reached an extraordinarily high level of civilisation, Lawrence insisted that the Egyptians were not Negroes but rather a "mixed population" composed mainly of Asian and European elements with a minor addition of the African race (*Lectures* 338–42). I mention this because Shelley's text twice identifies the Creature's skin with that of a "mummy" (*F* 53, 216). The only mummies that Shelley could have seen, either in the British Museum (where mummies were unwrapped and displayed as early as the 1790s) or when she visited the Louvre in 1814 (which housed the mummies brought back from Napoleon's Egyptian expeditions), were typically painted with faces ranging in color from pale yellow to reddish yellow to dark brown-yellow and, if unwrapped, possessed an embalmed yellow skin. Thus, this detail may further link the Creature with the Asian or Mongol race. Second, Lawrence, citing Peter Simon Pallas, insisted that the Mongols were a beardless race (*Lectures* 315). As Londa Schiebinger has noted, by 1815 the beard as a marker of male virility, and hence of both racial and sexual superiority, was widely assigned primarily to the Caucasian race.[17]

Percy Shelley probably first met William Lawrence in 1811, when he attended Dr. John Abernethy's lectures in anatomy at St. Bartholomew's Hospital with his cousin John Grove; certainly Lawrence had become Percy and Mary Shelley's personal physician by 1814, and Mary Shelley continued to consult him on medical matters and to meet him socially until his death in 1830.[18] Percy Shelley would have been particularly interested in the ideas of the intelligent, well-educated, and politically radical "Surgeon" Lawrence who repeatedly attacked the social inequities and intellectual bigotry of his day, even as he espoused his own racialist theories. That Percy Shelley had either read Blumenbach, or learned of his racial classifications from Lawrence, Prichard, Kames, or other natural historians, is suggested by his "Ode to the West Wind." There, in the opening lines, Percy Shelley hails the West Wind as the "breath of Autumn's being"

> from whose unseen presence the leaves dead
> Are driven, like ghosts from an enchanter fleeing,
> Yellow, and black, and pale, and hectic red,
> Pestilence-stricken multitudes. (lines 2–5)[19]

Numerous Shelley scholars have followed G. M. Mathews in identifying these pestilence-stricken multitudes of leaves, "yellow, and black, and pale, and hectic red," not only with the souls of the dead but specifically with the four races of man—"Mongoloid, Negroid, Caucasian and American Indian," as Donald Reiman has put it.[20]

That Mary Shelley was familiar with Blumenbach's racial categories is suggested in *Frankenstein* both by the Creature and by Victor. The Creature learns history by eavesdropping on Felix De Lacey as he reads Comte de Volney's *Ruins* (1791) to Safie. There Volney's Mahometans accuse the European Christians of attempting to destroy all other nations and races. I quote Volney's Muslims in a passage that Victor Frankenstein had echoed in an earlier section of Shelley's novel:

Was it the charity of your gospel that led you to exterminate whole nations in America, and to destroy the empires of Mexico and Peru; that makes you still desolate Africa, the inhabitants of which you sell like cattle, notwithstanding the abolition of slavery that you pretend your religion has effected; that makes you ravage India whose domains you usurp; in short, is it charity that has prompted you for three centuries past to disturb the peaceable inhabitants of three continents, the most prudent of whom, those of Japan and China, have been constrained to banish you from their country, that they might escape your chains and recover their domestic tranquillity?[21]

The Creature then summarizes Volney: "I heard of the slothful Asiatics; of the stupendous genius and mental activity of the Grecians; of the wars and wonderful virtue of the early Romans—of their subsequent degeneration— . . . I heard of the discovery of the American hemisphere, and wept with Safie over the hapless fate of its original inhabitants" (*F* 115).

By 1815 the image of the Mongols or Asians as a yellow-skinned, black-haired, and beardless race was well established, not only in the scientific writings of Blumenbach, Prichard, and Lawrence, but also in European culture at large. Starting in the 1780s, descriptions of the "yellow people" (who as Robert Markley has noted had previously been portrayed on maps and in paintings as white skinned) were disseminated by the ever-growing East India Company's trade with Asia and the Far East and appeared in diplomatic reports, missionary reports from India and China, the reports of the Asiatic Society of Bengal from 1790 onward, and such contemporary travel writing as John Barrow's immensely popular 1804 *Travels in China*, which included in the second edition (1806) a colored frontispiece portrait of the distinctively yellow-skinned Chinese ruler Van-ta-gin.[22]

Shelley's *Frankenstein* of 1818 initiates a new version of this Yellow Man, the image of the Mongol as a giant. Recall the only surviving visual representation of her Creature that Shelley is certain to have seen, the frontispiece to the revised edition (1831) of *Frankenstein*, designed by T. Holst and engraved by W. Chevalier (Figure 6.2), which you must imagine with yellow skin. In Shelley's novel, this gigantic yellow man is portrayed as a creature of superhuman strength and endurance, of intelligence and sensibility, a man who—denied the female companionship and family he so desperately craves—finally becomes a murdering monster, destroying all those dear to his maker.

Since the publication of this frontispiece in the 1831 edition of Shelley's novel, the image of the yellow man as a huge, degenerate monster has had a long and nefarious cultural life. By the 1880s the gigantic yellow man had become a synecdoche for the population of China as a whole, a population so enormous that it could, if mobilized, easily conquer all of Asia and Europe, as Genghis Khan had almost done six hundred years before. These hordes of yellow men soon became known as the "yellow peril," a term originally coined in German in 1895 as the title of Hermann

Figure 6.2. Mary Shelley, Frontispiece to *Frankenstein, or The Modern Prometheus* (London: H. Colburn and R. Bentley, 1831). Engraving by W. Chevalier from a drawing by T. Holst.

Knackfuss's allegorical painting, *Die Gelbe Gefahr*, based on a pencil design by the young Kaiser Wilhelm II himself (Figure 6.3). Here we see the "Deutsche Michael," the German Saint Michael, prepared to defend his Caucasian sister nations against the threatening arrival from the East of a yellow Buddha seated on a dragon in a storm cloud. The Kaiser gave this painting to the Tsar of Russia to signal Germany's shift from an alliance with Japan to an alliance with Russia.[23]

The British writer Matthew Phipps Shiel (1865–1947) first translated the term *gelbe Gefahr* into English in his xenophobic novel *The Yellow Danger* (1898) and promulgated it in his subsequent sensationalistic novels *The Yellow Wave* (1905) and *The Dragon* (1913), this last retitled *The Yellow Peril* in 1929. The fear of a "yellow invasion" was repeatedly aroused by the racist journal *The Wasp* in San Francisco in the late 1880s and 1890s to denounce the wave of cheap Chinese labor that flowed into California to work on the railroads and in the vineyards, laundries, and restaurants. Chinese immigrants were portrayed here visually as establishing a monopoly over certain trades (Figure 6.4) or worse, becoming a beacon or statue of liberty summoning ever more masses of opium-addicted and diseased settlers (Figure 6.5). The rabble-rousing cry against the "yellow peril" was a staple of American and British anti-Chinese and anti-Japanese propaganda during the 1904–1905 Russo–Japanese War.

This cry was taken up again in 1929 and sustained through the 1950s by the British author Arthur Henry Sarsfield Ward. Writing as Sax Rohmer, he created the demonic doctor Fu Manchu, who in numerous novels, and in the radio show and films based on these novels (Figure 6.6), aroused a powerful anti-Asian racial prejudice in yet another generation.[24] In his first novel, *The Insidious Doctor Fu-Manchu*, Rohmer described his famous protagonist as

> tall, lean and feline, high-shouldered, with a brow like Shakespeare and a face like Satan, a close-shaven skull, and long, magnetic eyes of true cat-green. Invest him with all the cruel cunning of an entire Eastern race, accumulated in one giant intellect, with all the resources, if you will, of a wealthy government which, however, already has denied all knowledge of his existence. Imagine that awful being, and you have a mental picture of Dr. Fu Manchu, the yellow peril incarnate in one man.[25]

But it was during the Second World War that Shelley's image of a gigantic, terrifying Yellow Man came to full visual fruition. Numerous anti-Japanese propaganda posters promoted the stereotype of the Japanese warrior as a gigantic bloodthirsty monster (Figure 6.7), barely distinguishable from an ape (Figure 6.8). In two posters designed in 1942 for a "This Is the Enemy" exhibition in New York City, the Japanese male was explicitly identified as an insidious sexual threat to defenseless white Caucasian females (Figures 6.9 and 6.10). This last poster uncannily echoes the Creature's murder of Elizabeth Frankenstein on her wedding night and takes me back to Shelley's novel.

Figure 6.3. Hermann Knackfuss, *Die Gelbe Gefahr* [The Yellow Peril], 1895, Drawing, from a design by His Majesty William II, German Emperor, King of Prussia. Reprinted from Arthur Diosy, *The New Far East* (London: Cassell, 1898).

Figure 6.4. G. F. Keller, *Allee samee 'Melican Man Monopoleeee, The Wasp* (20 May 1881): 32. Reprinted in P. Choy and L. Dong, *The Coming Man* (Seattle: U of Washington P, 1995), 91. Reproduced by permission of the Chinese Historical Society of America.

Figure 6.5. G. F. Keller, *A Statue for Our Harbor, The Wasp* (11 November 1881): 320. Reprinted in *The Coming Man* 136. Reproduced by permission of the Chinese Historical Society of America.

Figure 6.6. Film Poster for *Drums of Fu Manchu*, 1940. Reproduced by permission of Trentham Books Limited.

Figure 6.7. Carey Orr, *The Heroic Role*, January 1942, *Chicago Tribune.* © Chicago Tribune Company. All rights reserved used with permission.

A British commentary on the Japanese soldier

How Tough Are the Japanese?

They are not tougher than other soldiers, says a veteran
observer, but brutality is part of their fighting equipment.

Figure 6.8. Leslie Illingworth, *How Tough Are the Japanese?*, 1943, editorial cartoon for the *London Daily Mail*. Reprinted in *New York Times Magazine*, May 2, 1943, p. 6. © *The Daily Mail*; reproduced by permission.

Figure 6.9. H. Melzian, *This Is the Enemy*, 21 December 1942, *Life Magazine* (55); submitted to a "This is the enemy" war-poster competition and exhibited at the Museum of Modern Art, New York, November, 1942–January, 1943.

Figure 6.10. G. V. Lewis, *This Is the Enemy*, 21 December 1942, *Life Magazine* (55); submitted to a "This is the enemy" war-poster competition and exhibited at the Museum of Modern Art, New York, November, 1942–January 1943.

Does the Creature's yellow skin—together with the animal as well as human parts from which he is constructed—indicate that he is by his very bodily nature a degenerate being, both racially and evolutionarily inferior to his Caucasian creator, and hence necessarily a monster? Before I define Mary Shelley as a powerful contributor to that nineteenth-century British racist science so well analyzed by Nancy Stepan and George Mosse, we should look at her novel very closely.[26] Rather than condemning Frankenstein's creature as a member of a morally degenerate race, I would argue, Mary Shelley encodes in her novel a possible *solution* to racial stereotyping and racial hatred. Throughout her novel, the Creature has asked to be included within a nuclear family, either to belong to an already established Caucasian family unit or to be given a wife and children of his own kind whom he can cherish. He first tries to ingratiate himself with the French De Lacey family and is accepted by the blind father, only to be rejected by the terrified children. He then tries to adopt little William Frankenstein but accidentally strangles the child in his unwittingly strong embrace. Finally, he begs Victor Frankenstein to create a mate for him, a female creature who will be the Eve to his Adam and accompany him to the wilds of the new world, of South America, to procreate his own race. Only when Victor Frankenstein refuses to provide a family for his creature, does the Creature become vengeful and violent.

Let us return to the published frontispiece of the 1831 *Frankenstein* (see Figure 6.2). Victor Frankenstein tells us that he had selected the Creature's form and features as "beautiful"—and indeed, in this frontispiece, the Creature does appear with a huge but perfectly proportioned classical European nude body. Only the head with its possibly Mongol features appears awkwardly connected to the body. What we are seeing, I suggest, may be a racial hybrid—a Caucasian body with Mongolian yellow skin, eye color, hair, and beardlessness. To Victor Frankenstein's eyes, such a racial hybrid can be only a degenerate monster. But to Shelley's eyes, such a racial "amalgamation" (to use the term Shelley would have known) might represent a positive evolution of the human species.[27] Shelley implicitly raises the possibility of a future mating between the white and the yellow races in the scene in which Victor Frankenstein refuses to continue making a female creature. Among other rationalizations, Victor is terrified that this gigantic yellow female will procreate a "race of devils" and even worse, "might turn with disgust from [the Creature] to the superior beauty of man" and prefer to mate with a white man, perhaps even with Victor himself (*F* 163). Victor can see such interracial mating only as a repulsive miscegenation; in horror, he rips up the female creature.

But what if we as readers were to see such interracial mating as a positive act, an act that embraces every race within a single Family of Man? If we did, then we might see Shelley as aligning herself with several other British women writers of the early nineteenth century who represented interracial mating in a positive, even a romantic light. I am thinking of Felicia Hemans, the leading female poet of the

Romantic era who implicitly endorsed the sexual love between European Christians and Moorish Muslims in her *Siege of Valencia* (1816); of Hannah More, who staged the love of the Roman Publius for the African slave girl Barce as entirely noble in her play *The Inflexible Captive* (1774); and of Esme Erskine (Eliza Bland Norton), who in her epic poem *Alcon Malanzore* (1815) hailed the marriage between a Spanish Catholic woman and an African Muslim man as the solution to religious warfare. Shelley may also be in accordance on this point with her husband, who soon after in his epic drama, *Prometheus Unbound* (1820), celebrated the marriage of the Greek god Prometheus, here confined to the Caucasus, with the eponymously named Asia, who represents both the land and the race east of the Caucasus/Europe. Rather than subscribing to Lawrence's racialist hierarchies, Mary Shelley may be suggesting that racial difference and interracial mating are social evils *only* when we see them and write them as evil. Recall that in *Frankenstein*, Felix De Lacey and his family eagerly welcome the Arabian Safie into their European family.

In her next novel, *The Last Man* (1826), Shelley further implies that only by embracing another race can one escape that fatal plague which threatens to annihilate the human species. She first identifies the plague—cholera or the "yellow fever"—as geographically and racially "foreign," originating in Africa and then spreading throughout Asia. Initially, the Caucasians assume that they are racially immune to this disease: "It drinks the dark blood of the inhabitant of the south, but it never feasts on the pale-faced Celt."[28] But in *The Last Man*, no race is immune: the plague easily infects the "healthy constitutions" of Europe.

And once infected, no one recovers from the plague—*except Lionel Verney*. Verney's survival suggests a possible alternative to this tale of universal destruction. Recall that Verney succumbs to the plague when, hearing a moan, he compassionately enters a dark room where he is "clasped" by "a negro half clad, writhing under the agony of the disease": "I strove to disengage myself, and fell on the sufferer; he wound his naked festering arms round me, his face was close to mine, and his breath, death-laden, entered my vitals" (245). From this unwilling but powerful embrace of the racial other (this is the only time that a "negro" is mentioned in *The Last Man*), Verney both contracts and, recovering, becomes immune to the plague.

Does this episode imply that if one were to embrace the racial other, however unwillingly, rather than to define it as "foreign," "diseased," "degenerate," or "monstrous," then one might survive the plague? If so, then *The Last Man* supports the suggestion I am making about the function of race in *Frankenstein*. Throughout *The Last Man*, Shelley suggests that the plague is a socially constructed as much as a natural disease. It is thus a parallel to the creation of the monster in Frankenstein: a monster who—as I have argued at length elsewhere[29]—is not innately evil but is rather the product of the textual readings imposed on him by the paranoid imaginations of his fellow creatures. If Victor had been able to see his gigantic yellow-skinned Creature as a member of the Family of Man, if he had been able to provide him with the domestic affection he craved—as does the couple in Diane Arbus's famous photograph of *The Jewish giant at home with his parents in the Bronx, 1970*[30]—then he might indeed have produced a more highly evolved variety of the human species–one capable of advancing rather than destroying human civilization.

Notes

I am grateful to my colleagues Benjamin A. Elman, Jinqi Ling, and Miriam Silverberg, and to Christopher Keep for their invaluable help with this chapter.

1. Mary Wollstonecraft Shelley, *Frankenstein or The Modern Prometheus*, ed. James Rieger (1974; Chicago: U of Chicago P, 1982), 52. Cited parenthetically in the text hereafter as *F*.

2. A few scholars have called attention to Victor Frankenstein and Clerval as racial imperialists (see Gayatri Chakravorty Spivak, "Three Women's Texts and a Critique of Imperialism," *Critical Inquiry* 12 [1985]: 243–61; Joseph W. Lew, "The Deceptive Other: Mary Shelley's Critique of Orientalism in *Frankenstein*," *Studies in Romanticism* 30 [1991]: 255–83); and some have noted the racial "otherness" of Frankenstein's creature (see Zohreh T. Sullivan, "Race, Gender and Imperial Ideology in the Nineteenth Century," *Nineteenth-Century Contexts* 13 [1989]: 19–32; Harold L. Malchow, "Frankenstein's Monster and Images of Race in Nineteenth-Century Britain," *Past and Present* 139 [1993]: 90–130; and Malchow's *Gothic Images of Race in Nineteenth-Century Britain* [Stanford: Stanford UP, 1996]; but none have accurately defined the Creature's race in the context of nineteenth-century racial science. Spivak and Lew rightly associate Victor Frankenstein and Clerval with the orientalizing imperialism of the East India Company. Sullivan suggests that the Creature may be "Burmese," while Malchow—on the basis of no textual evidence—insists that he is African.

3. Anne K. Mellor, " *Frankenstein*: A Feminist Critique of Science," *One Culture: Essays in Science and Literature*, ed. George Levine (Madison: U of Wisconsin P, 1987), 287–312; and Mellor, *Mary Shelley: Her Life, Her Fiction, Her Monsters* (London: Methuen/Routledge, 1988).

4. Georges-Louis LeClerc, Comte de Buffon, *The System of Natural History*, abridged, 4 vols. (Alnwick, UK: Printed for W. Davidson, 1814), 107. In this translation of Buffon's *Histoire naturelle*, his discussion of the "Apparent Varieties in the Human Species" appears in vol. 1, ch. 6, 107–30; cf. John Herbert Eddy Jr., *Buffon, Organic Change, and the Ages of Man* (Ann Arbor: University Microfilms International, 1977), 107–48.

5. Nicholas Hudson, "From 'Nation' to 'Race': The Origin of Racial Classification in Eighteenth-Century Thought," *Eighteenth-Century Studies* 29 (1996): 247–64; Ivan Hannaford, *Race—The History of an Idea in the West* (Baltimore: Johns Hopkins UP, 1996). As Hudson and Hannaford show, the concept of race (as opposed to nation or tribe) was a late-eighteenth-century construction, loosely based on the divisions between the five known continents. The first person to use the term "race" in its modern sense was François Bernier, who in 1684 identified four distinct human races. See his "Nouvelle division de la terre, par les differentes especes ou races d'hommes qui l'habitent . . . ," *Journal des savants* 12 (1684): 148–55.

6. Johann Friedrich Blumenbach, *On the Natural History of Mankind*, in *The Anthropological Treatises of Johann Friedrich Blumenbach*, trans. Thomas Bendyshe (London: Longman, 1865), 65–145 (1775 text), 145–278 (1795 text). Cited parenthetically in the text hereafter as *Mankind*. Cf. *Contributions to Natural History* (1806), ibid., 277–340; *The Elements of Physiology*, trans. Charles Caldwell (Philadelphia: Thomas Dobson, 1795); and *A Short System of Comparative Anatomy*, trans. William Lawrence [with additional notes and an introductory view of the classification of animals by the translator] (London: Longman, 1807).

7. On Charles White's *An Account of the Regular Gradation in Man and in Different Animals and Vegetables* (London: Printed for C. Dilly, 1799), see Miriam Claude Meijer, "The Anthropology of Petrus Camper (1722–1789)" diss., U of California–Los Angeles, 1991.

8. Henry Home, Lord Kames, *Sketches of the History of Man* (Edinburgh: Printed for William Creech, 1774); 2nd ed. (1778); 3rd ("considerably improved") ed., 2 vols. (Dublin: James Williams, 1779).

9. *An Essay on the Causes of the Variety of Complexion and Figure in the Human Species* (London: Printed for John Stockdale, 1789). A reissue of the "enlarged and improved" second edition, edited by Winthrop D. Jordan, appeared in 1965 (Cambridge, MA: Harvard UP).

10. *Researches into the Physical History of Man*, ed. George W. Stocking Jr. (Chicago: U of Chicago P, 1973), 1, 22.

11. Ibid., iii.

12. Marilyn Butler, "Introduction," *Mary Shelley's 1818* Frankenstein (Oxford: Oxford World's Classics, 1994), i–xx.

13. *Lectures on Physiology, Zoology, and the Natural History of Man* (London: J. Callow, 1819), 74. Cited parenthetically in the text hereafter as *Lectures*.

14. See Buffon, *Natural History* 1:107–30.

15. Cf. Edward W. Said, *Orientalism* (New York: Random House, 1978).

16. James Chambers, *The Devil's Horsemen—The Mongol Invasion of Europe* (London: Weidenfeld and Nicolson, 1979; rev. and extended, Cassell, 1988).

17. *Nature's Body—Gender in the Making of Modern Science* (Boston: Beacon, 1993), 120–25.

18. Hugh J. Luke Jr., "Sir William Lawrence—Physician to Shelley and Mary," *Papers on English Language and Literature* 1 (1965): 141–52; cf. Mary Shelley, *The Journals of Mary Shelley: 1814–1844*, 2 vols., ed. Paula R. Feldman and Diana Scott-Kilvert (Oxford: Clarendon, 1987), 1:55, 67 n.1, 180 n.8; 2:512; Mary Shelley, *The Letters of Mary Wollstonecraft Shelley*, 3 vols., ed. Betty T. Bennett (Baltimore: Johns Hopkins UP, 1980–1988), 1:41; 2:111, 210; 3:84. Luke's account is corrected by Donald Reiman, *Shelley and His Circle: 1773–1822*, vol. 5 (Cambridge, MA: Harvard UP, 1973), 44.

19. Percy Bysshe Shelley, "Ode to the West Wind," *Shelley's Poetry and Prose*, ed. Donald H. Reiman with Sharon B. Powers (New York: Norton, 1977), 221.

20. Ibid., 221 n.3.

21. Constantine Francis Chasseboeuf de Volney, *The Ruins: or, A Survey of the Revolutions of Empires*, first English translation, 3rd. ed. (London: J. Johnson, 1796), 307–08, ch. 23.

22. See, respectively, Robert Markley, "Civility, Ceremony, and Desire at Beijing: Sensibility and the European Quest for 'Free Trade' with China in the Late Seventeenth Century," *Passionate Encounters in a Time of Sensibility*, ed. Maximilian E. Novak and

Anne Mellor (Newark: U of Delaware P, 2000), 60–88; *Asiatick Researches: or, Transactions of the Society instituted in Bengal, for inquiring into the history and antiquities, the arts, sciences, and literature of Asia* (Calcutta: Printed and sold by Manuel Cantopher, 1788–1839), see especially vol. 2 (1790); and John Barrow, *Travels in China* (1804; London: Cadell and Davies, 1806).

23. Arthur Diosy, *The New Far East* (1898; London: Cassell, 1904), 327–36; V. G. Kiernan, *The Lords of Human Kind—Black Man, Yellow Man, and White Man in an Age of Empire* (Boston: Little, Brown, 1969), 170–72.

24. Jenny Clegg, *Fu Manchu and the Yellow Peril* (New York: Trentham, 1994); William F. Wu, *The Yellow Peril—Chinese Americans in American Fiction, 1850–1940* (Hamden, CT: Archon, 1982), Ch. 6.

25. Sax Rohmer, *The Insidious Doctor Fu-Manchu* (1913; New York: Pyramid, 1961), 17.

26. Nancy Stepan, *The Idea of Race in Science: Great Britain, 1800–1900* (London: Macmillan, 1982); George L. Mosse, *Toward the Final Solution—A History of European Racism* (New York: Howard Fertig, 1978).

27. Robert J. C. Young, *Colonial Desire: Hybridity in Theory, Culture, and Race* (New York: Routledge, 1995), 9.

28. *The Last Man*, ed. Hugh J. Luke Jr. (Lincoln: U of Nebraska P, 1965), 245.

29. Mellor, *Mary Shelley*.

30. This image could not be reproduced here for copyright reasons.

Part III

*Botany, Taxonomy,
and Political Discourse*

CHAPTER SEVEN

Lyrical Strategies, Didactic Intent

Reading the Kitchen Garden Manual

RACHEL CRAWFORD

By the end of the Regency period in 1820, an era when British imperial fortunes seemed boundless, there emerged a new sense of the kinds of spaces productive of Englishness, one that differed profoundly from the boundless prospect espoused early in the eighteenth century. These were the vernacular spaces of the cottage garden, the homes and hearths of the middle class, and, in literature, minor lyric forms that materialized in dizzying numbers in popular magazines, correspondence, and parlor games, and that were the subject of impassioned critical exchange. Contained spaces geared toward productivity (intellectual, patriotic, domestic, and horticultural) usually, although not uniformly, of smaller dimension, took hold of the English imagination. Such sites attracted the attention of the architects of space: building designers, gardeners, agriculturists, and poets. Susan Stewart has suggested, "we can posit an isomorphism between changes in genre and changes in other modes of production."[1] The congruencies Stewart postulates—relationships between things drawn by virtue of their forms—may be better understood in terms of shared social proportion and weight even when materials (e.g., words and space) may display different properties. Eighteenth-century texts reveal such congruencies and situate changes in the concept of social space in the interstices of topography and literature. Indeed, although Stewart cautions that "historical and generic conventions cannot be mapped upon the real," the authors of eighteenth-century treatises that mediate the "real" practices of gardening and poetic composition pilfer from each other's disciplines as they work to build consensus and authority for their respective trades. They readily negate material differences between words and space in order to privilege social and theoretical congruencies. In their minds, it would seem, though the boundaries between landscape and literature may be posted, they are negotiable by the hucksters of language.

In this essay, I focus on changes in the concept of social space that reconstituted notions of containment in the latter decades of the eighteenth century and on social and scientific values that informed these changes. The aesthetic that emerged from the fascination with the relationship between containment and productivity was influenced by kitchen garden design and plant taxonomy, helped to shape requirements for lyric verse, and naturalized reproduction as the telos of female sexuality. It included within its parameters the formal requirement of visible artifice, the affective requirement of sublime power compressed within contracted sites (rather than set into play by the boundless prospect), the synaesthetic effect of temporal diffusion, and an inherited erotic discourse that was habituated to the contained, contrived space. This aesthetic assisted in raising lyric poetry from the bottom of eighteenth-century poetic taxonomies to the elevated position it still retains and provided vernacular sites such as the kitchen garden with representational status in a national agenda.

The critical discourses surrounding kitchen gardens and lyric verse consistently share an assumption that productivity is best achieved within confined space. Unlike literary critics, however, authors of kitchen garden manuals from the first decades of the eighteenth century extolled the virtues of containment. By the conclusion of the eighteenth century proper when lyric poetry had acquired unaccustomed prestige on the scale of literary forms, an aesthetic of exuberance within confined space and of productivity linked to functionality would supplant that of the great park with its boundless view. This aesthetic established representative values of Englishness over the course of the nineteenth century that remain powerful today. The critical discourses that constituted these literary and topographical ideals provided cultural intelligibility for values that informed notions of middle-class space, leisure, plenty, and taste. Sexual norms also entered the nexus of values created by the convergence of containment and productivity in part because an archaic, inherited set of assumptions correlate confined sites with female sexuality, in part because empirical information about the female reproductive system provided women with ontological status.[2] Congruent with these values, natural taxonomies in botany, which classified plants by visible, often morphological features, shifted temporarily to Linnaeus's sexual system, which limited classification to reproductive components. Sexuality in the garden, as in female reproductive function, linked delight to telos; and both expressed the link between containment and productivity extolled by promoters of lyric verse.

The Kitchen Garden Manual

The view that beauty could be geared toward productivity, that it was a function of use, had been pressed by designers of kitchen gardens for the entire century and legitimated by Kames in 1762 as relative beauty, or beauty "relating to some good end or purpose."[3] Enthusiastic about the re-creative prospects of this small space, authors of kitchen garden manuals found precedent in the bowers of Milton's Eden and Spenser's *Fairie Queene* and ostentatiously borrowed their vocabulary from lyric passages in the fertile tradition of English poetry. It is hardly odd, therefore, that descriptions in manuals and calendars of gardens brimming with fruits, vegetables, flowers, and herbs should exhibit techniques of lyric form while employing didactic strategies. Indeed, manuals provide a singular continuity between those decades delighted by the expansive reaches of English georgic poetry and the post–American War fascination with minor lyric forms. Unlike georgic poetry, in which the embowered space is hidden within expansive prospects and subsumed by purposive instruction, kitchen garden manuals extract the small enclosed space from the larger text of the landscape and in so doing unite didactic purpose with lyrical technique, *utilis* with *dulci*. Treatises on the kitchen garden contributed to both georgic and lyric impulses, conforming to the profitability, industry, and occasion for instruction required by the former and to the artifice and sublime immediacy of the latter. Through them we glimpse a capacious public mentality that valued both the curious and the visceral, both instruction and profligacy. In the continuous tradition of the kitchen garden manual we watch an aesthetic for contained space taking shape and discern a convergence between topographical and literary form.

A growing interest in the healthful qualities of vegetables, "physical" properties of herbs, and commercial possibilities of flowers combined with new enthusiasm for dietary variation was already well established early in the century. In addition, as the long tradition of labor-intensive farming in the southwest Midlands proved, the methods practiced in the kitchen garden had profound implications for the practice of agriculture, providing methods for growing food that could boost the productivity of farmlands. Richard Weston and Humphry Repton were only two of many gardeners who advocated transferring methods from kitchen gardening to farming.[4] Their comments feed into a growing sense that containment is a precondition for productivity and, like those on both sides of the heated contemporary parliamentary enclosure debates, link containment to a national purpose. Writers of these manuals are also convinced that a properly managed kitchen garden provides pleasure and beauty. Unlike Weston, who calculated the profit in roses, peppermint, and poppies and recommended planting onions around the base of fruit trees because they guarantee income their first year, others marveled openly at the beauties of this vernacular space. The numerous manuals devoted to garden practice and design reveal that because beauty in the kitchen garden was viewed as an effect of its function, these gardens provided a way of defining the elements of form for a contained space quite distinct from those of the landscaped park. The functional aesthetic, obscured by the passion for the "natural" contours of the English landscape garden, was brought by means of the kitchen garden into the domain of pleasure.

The convergence of aesthetic requirements for artifice, sublimity, and temporal diffusion in the discourses surrounding kitchen gardens and minor lyric forms
is expressed most clearly in prefaces and essays on lyric poetry that began to pro-
liferate during the 1780s. Literary critics and poets often take their cue from the
poet Edward Young and adapt his principles for the great ode to lesser lyric forms
such as the Horatian ode and sonnet, which they treat as serious poetic forms. The
visible transversion between critical texts and garden manuals underscores the fact
that, beyond expressing a subjective sensibility, minor lyric was perceived as a formal *space* that generates content. Its brevity, indeed, as W. R. Johnson points out,
was not viewed simply as beautiful but became the standard for beauty.[5] Literary
critics and authors of garden manuals share assumptions concerning artifice, sublimity, and temporal diffusion that are historically specific and socially weighted.
In particular, despite the fact that lyric is a self-reflexive form while garden manuals are instructive, both fabricate the idea of a self-enclosed world in which the
speaker whispers poetry's secrets into her own ear. In kitchen garden manuals the
self-enclosed world is, like Milton's Eden, a horticultural plenum filled with diverse
fruits and vegetables from the four corners of a Britannizable world. Most strikingly, lyric's self-reflexivity, its self-proclaimed enunciation of private emotion, is
ironically designed for the highly public purpose of distribution and circulation.
Inversely, as a didactic text the kitchen garden manual opens with the assumption
that the speaker occupies a public forum, yet presumes, as the etymology of the
word *garden* suggests, a guarded and secluded space—one which is shielded from
the eye of the passerby and shuts out the external prospect. As readers of the kitchen
garden manual we figuratively peer through the chink in the wall or the break in
the hedge into our neighbor's yard. The allure in each case, so unlike the indiffer-
ent gaze over the wide prospect presumed by the aesthetic of the landscaped park
or open forms of georgic and the great ode, is the allure of the bounded, the
excluded, the voyeuristic gaze.

Manuals devoted to the kitchen garden reveal that confined space is conceived
around fundamentally different aesthetic principles than boundless space and incorporate a logic that privileges the lyric strategies of compression, delineation, and
contrivance. Throughout the eighteenth century, authors of garden manuals and
landscape treatises maintained that the contained space requires its own set of aesthetic guidelines—that it could not without absurdity simply miniaturize larger
forms.[6] The constriction of space could not be understood simply as a syncope of
the open prospect. Authors of garden manuals project an aesthetic fitted to the differing demands of a contained space which, designed to be useful, was
enthusiastically acclaimed for its beauty. In drawing attention to the qualities of artifice, sublimity, and temporal diffusion and the strategies of compression,
delineation, and contrivance, they formulate assumptions shared by commentators
on lyric verse. That is to say, although lyric verse and didactic prose texts presume

different properties of voice and composition, they share historically specific assumptions, social weight, and aesthetic concerns.

From the early years of concerted fruit and vegetable gardening, writers of garden manuals were engaged in plotting the intersection of garden design and productivity. Batty Langley's *New Principles of Gardening* (1728) testifies to the porous boundaries between notions of design (which promote pleasure) and the cultivation of fruits and vegetables (which promote utility and profit) that existed throughout the century, but which have been obscured to modern readers by the flood of treatises describing great parks or, in the last quarter of the century, regulating the taste for the picturesque. In Langley's manual, principles of design indebted to Addison infiltrate and abet principles of cultivation. "The End and Design of a good Garden," he says, "is to be both profitable and delightful; wherein should be observed, that its Parts should be always presenting new Objects, which is a continual Entertainment to the Eye, and raises a Pleasure of the Imagination."[7] His ideal garden is a tumult of components, an entire world in which pleasure and utility jostle together. Likewise, Philip Miller prefaced his encyclopedic manual, editions of which spanned the eighteenth century, with a series of rhetorical questions that advance the kitchen garden's claims for delight and use:

> What can be more *delightful*, than in the *Spring-time*, to behold the Infant Plants putting forth their verdant Heads, from the Bosom of their fostering Mother the earth? In the *Summer Months*, the Flowers ting'd with a Variety of the most charming Dyes, seeming, as it were, to vie with each other, which shall most allure the Beholder's eye with their splendid Gaiety, and entertain the Nostrils with their enlivening Fragrancy? and in *Autumn*, to view the bending Boughs, as it were submissively offering their delicious Fruit, and courting the Gatherer's Acceptance?[8]

Similarly, little less than a century after Langley, landscape gardener Humphry Repton espoused the principle that beauty is one of the products of use. In his Redbook for Blenden Hall, he disparages the fashion for vast lawns and, more unexpectedly, for agriculture, advocating instead the beauty of the garden. In his view, "it is not necessary that the garden should be a square area within four walls. A fruit garden may be so blended with flowers and vegetables, as to be interesting in all seasons; and the delight of a garden highly cultivated, and neatly kept, is amongst the purest pleasures which man can enjoy on earth."[9] Repton's Regency era texts encourage an Edenic ideal at the same time that they whet the appetite. Beauty and productivity are met together in a visual metaphor of taste that is both aesthetic and appetitive in the kind of garden which these writers believe is truly English. Treatises on parks, by contrast, even such lively and amusing histories as Horace Walpole's "On Modern Gardening" (1780) or Sir William Chambers's outrageous *Dissertation on Oriental Gardening* (1772), appealed almost exclusively to sight and the psychological effects of the categories of the landscape. The gustatory temptation proffered by garden manuals produces a didactic mode that joins the pleasures of the body with those of the mind. Artifice, sublime profusion, and temporal diffusion meet in a delimited, contrived space that, among its precedents, looks to British poetry.

In the kitchen garden manual, lyric principles are separable only arbitrarily. To examine one is to open a window on each of the others. The role of the gardener demonstrates particularly well how interinvolved these principles become. It had been a conceit since Francis Bacon's essay "Of Gardens" (1597) that "God almighty first planted a Garden" and that Adam was therefore, in Philip Miller's words, a "Master Gardener."[10] The role of the gardener is therefore consistently defined as "recreative," even when the owner and actual gardener are not clearly distinguished. In calendar entries, especially, the gardener is portrayed as both regulator and genius of the garden. Under his tutelage, the garden is a product of artifice in which design, although everywhere apparent, produces sublime disorder. Thus Stephen Switzer, following John Evelyn's translation of a garden manual by Jean de la Quintinye, director of the royal gardens in France, describes the garden in June as though it were a parterre in a park:

> All the squares of the garden are now cover'd with green herbs, which compleats that natural tapestry with which the ground is or ought to be adorn'd; we gather, in all parts of the garden, such things as are ready and proper for it; and at the same time, with an agreeable profusion, distribute all those plants that are become so beautiful and accomplish'd as to fill up other places, which we now do, so that there hardly ever remains any part of space of our garden void.[11]

Purposeful art is essential to the kitchen garden's rampant productivity since Nature is inadequate to these productions, either withdrawing her aid or hampered by England's "variable and uncertain" climate.[12] Within the kitchen garden the effects of the Fall, especially the withdrawal of Nature's *desire* to assist humankind, can be reversed by the diligence of an artful gardener, and the moral attributes that led to the Fall can be redefined through ingenuity.

Switzer reveals the intertwined discursive traditions of poetry and garden manuals when he compares the fruit garden to our original paradise, presuming reader familiarity with *Paradise Lost:*

> And indeed, if Fruit-Trees had no other Advantage attending them than to look upon them, how pleasurable would *that* be? Since there is no flowering Shrub excells, if equals, that of a Peach or Apple-Tree in Bloom. The tender enammell'd Blossoms, verdant Foliage, with such a glorious Embroider of Festoons and Fruitages, wafting their Odours on every Blast of Wind: And at last bowing down their laden Branches, ready to yield their pregnant Offspring into the Hands of their laborious Planter and Owner. Indeed a well contriv'd Fruit-Garden is an

> Epitomy of Paradise it self, where the Mind of Man is in its highest Raptures, and where the Souls of the Virtuous enjoy the utmost Pleasures they are susceptible of in this sublunary State. For there the happy Planter is cooling and refreshing himself with *Scooping the brimming Stream* of those nectarous Juices, and the philosophizing thereon, as Mr. Milton has it in that excellent Description, Book iv.l.327. of *Paradise Lost;* also Mr. *Philips* very rapturously describes it in his Poem on *Cyder.*[13]

This passage begins with a rhetorical question that suggests that the purpose of the fruit tree is not its beauty; Switzer nevertheless proceeds to the scene of productivity by describing the beauties of orchard flowers in the highly suggestive language of romance. In the most familiar terms of that tradition, Englished by Spenser, polished by Sidney, and domesticated by Milton, the language of the garden is both artful and sexual. Strewn with enameled blossoms and embroideries of flowers and fruits, its space is explicitly the product of contrivance, and for Switzer it follows that it is the epitome of paradise. The scene is sensualized by images of tender blossoms, verdant foliage, wafting odors, pregnant offspring, mental raptures, and utmost pleasures assisted by brimming streams and nectarous juices. Significantly, Switzer substitutes *in*geniousness for Genius, warden of Spenser's Garden of Adonis. In this garden one eats the fruit of the tree grown through the knowledge not only of good but of art.

The goal of artifice in garden manuals is to re-create within the narrow compass of garden walls or even narrower confines of the greenhouse a world replete not only with innumerable indigenous fruits and vegetables but also with artificially cultivated produce of every latitude. This goal redirects the hope expressed by agriculturists that England would become the Garden of the World through a system of self-contained plots. In the latter view, England would radiate principles of agricultural practice onto less progressive nations. The image is expressive. In the kitchen garden manual the garden is a world in itself. The image is centripetal, of attraction toward a well-defined center from "outlandish" regions. In the English garden the outlandish or exotic are "endenizon'd," to use Switzer's word, making foods safe and healthful which might otherwise poison the social body.[14] Switzer provides a graphic narrative of the process in his account of the virtue of peaches. Ancient authors, he tells us, claimed "that the *Persians* from whence they were first brought, and from whence they deriv'd their Names, dar'd not eat of them by Reason of their Malignity." He informs us that even now,

> in *Persia* those Fruits have some malignant and over purgative Dispositions; but since those Trees were transported into *Egypt*, from thence replanted in *Italy*, and since that cultivated in *France* and other *European* Countries, they have lost those pernicious Qualities, and retain nothing but those that are purely purgative; and for this Virtue it is that they are esteem'd by the lovers of Health, who had rather eat a few Peaches in a Morning than take a Dose of Physick.

The progress of the peach through a series of countries presented as increasingly enlightened and, not incidentally, ever more proximate to England, filters the poisons out of the fruit. It thus reaches the apex of its healthful powers in England, where "the Art and Industry of our Gardeners have render'd [it] so commendable."[15]

The kitchen garden parallels in efficacy an imperial foreign policy in which alien peoples, like pineapples and oranges, melons, or peaches, could, carefully managed, be productively nurtured and safely absorbed within a recognizably British order. The wholly assimilated foreigner, like the ancient conquerors of England, *is* English, having improved the native stock, while becoming, in the process, improved. The authors of these treatises thus reveal the sense of native exuberance associated with bringing the world into the garden as well as the concomitant fear of the risks involved in the importation of foreignness within domestic boundaries.

The idea that the world could be brought into a garden is one filament in a web of discourses that support a view of England as Edenic because of its indigenous virtue and imperial success. The image is inseparable from Milton's depiction of Eden, in which we find "To all delight of human sense expos'd/In narrow room Nature's whole wealth, yea more,/A Heaven on Earth."[16] Rather than infinite space, heaven is emblematized as a walled garden planted with "All Trees of noblest kind for sight, smell, taste" (*Paradise Lost* 4:217). Its interior, like Blake's grain of sand, is larger than its exterior. But whereas Milton claimed that Eden was born from "not nice Art" but Nature (*Paradise Lost* 4:241–42), the kitchen garden required artifice of the highest sort and celebrated contrivance in the production of its fruits and vegetables. In fact, while the idea of the world contained in the garden is given authority by its Miltonic precedent, it takes hold primarily because the arts of the garden assist and are assisted by commerce.

Productivity, created by compression and motivated by the systolic impetus of commerce, drove descriptions of the contents of the kitchen garden and greenhouse. These display profuse growth and productive power within a contracted and sharply defined area and utility that generates extravagant beauty. For an audience schooled in didactic literature and still delighted by the georgic poem, garden manuals depicted the dizzying bounty of England's soil as both profligate and sensual. Within the contained space of the garden, one encounters sublime profusion. Although subject to the governance and ministration of a gardener, fruits and vegetables are depicted as lively, growing things:

> Here we see artichokes rising as it were from the dead; and there asparagus piercing the ground in a thousand places; here we should with pleasure observe cabbage lettuces wind themselves up into round balls; and there multitudes of legumes and green herbs, so different in colour, and so various in their shape, that a contemplative man can't but stand still with wonder and amazement; these! these! are the innocent and natural dainties, where they present themselves and grow for the nourishment and delicious entertainment of human kind.[17]

As this passage indicates, a small garden is the site of vast productivity. Containing walls and divisions abet its productivity, since it bursts with pears from every wall, cabbages and artichokes from every parterre. Descriptions of great parks, thin by comparison, are psychologically rather than viscerally oriented. Garden calendars, especially, established a pattern of heaping fruits and vegetables in lists that by their sheer magnitude suggest the potentiality of English soil. Such lists, with dozens of varieties contained within specific kinds, like epic catalogs combine excess with purposeful order.

Manuals and calendars take such techniques to marvelous ends, conjuring images of Edenic gardens that provide endless fruit, vegetables, and flowers. Some lists jumble all three together, especially within the confines of the greenhouse. Kidney beans, for example, rub shoulders with early dwarf peas, large lettuce plants, and strawberries in John Abercrombie's *The Hot-House Gardener*; these give way to pots of mint, tarragon, and tansey, which are followed in abrupt transition by dwarf tulips, hyacinths, jonquils, irises, anemones, pinks, roses, hypericum, and Persian lilacs.[18] Other lists produce a slightly different effect by cataloging many varieties of single fruits. A catalog of the twenty different kinds of "oranges" available by 1769 does more than provide a rational typology of plants. The English names crowded together conjure exotic fruits through which the mythologized countries of their origins materialize. Gnarled picturesque plants are embedded between more elegant specimens, and the hothouse is pervaded by its own strangely vegetative form of eros. Thus are listed "common Seville, sweet Seville, China, curl-leaved, striped curled-leaved, horned, common striped, hermaphrodite, willow-leaved or Turkey, striped Turkey, Pumpelmoes or shaddock, double flowered, common dwarf or nutmeg, dwarf striped, dwarf China, childing, distorted, large-warted, starry, sweet rinded."[19] Similarly teeming lists of varietals appear for lemons, citrons, pears, peaches, nectarines, plums, and, of course, England's own fruit, the apple. Repeatedly inserted et ceteras (*&c.*) in such lists suggest a plenitude the manual can strive to call forth, but can only intimate. The form of the manual, constricted in space and time, must always fail to capture the profligacy of nature assisted by art. Rhetorically, the compression of proliferated images, evocation of taste, odor, and form, and radical turns from one kind of produce to another, generate an effect which, while predominantly didactic, is also decidedly lyrical. Use becomes the foundation of beauty, a subtle shift in the relationship between the two that William Mason captures in his Horatian georgic, "The English Garden," when he proclaims, "*Beauty* scorns to dwell/Where *Use* is exil'd."[20] Pure didacticism—useful discourse from which every extraneous element has been boiled away—reveals itself as essentially lyrical and sublime.

As in the minor lyric, therefore, expansion from within is the paradoxical metaphor for temporality within containment. The anchoring image is Eden: a

place extracted from its larger context; a world compressed within the secure confines of the wall or hedge; a space within which fruit trees blossom and ripen and
vegetables grow and reach maturity simultaneously. Time in the kitchen garden is
a contained and rational infinitude, for by bringing the world into the garden the
gardener announces his power to control time by limiting space. The process of
mapping time onto space is especially vivid in the garden calendar. In these pocket-
size manuals the progress of time has a logical yet arbitrary linearity. The author
begins with January and ends with December. Reading from beginning to end,
however, the reader becomes aware that the experience of gardeners is quite different, since they encounter time as ritually circular—January and December the
neap tide in a seasonal cycle that achieves a plenum in July, August, and September. Even such a reordering of the calendar, however, does not catch the quality of
continuous production effected through the agency of the hothouse. In the "glassary compartments" of this world-within-a-world, seedtime and harvest occur
simultaneously: cucumbers are plucked in December and pineapples and oranges
ripen as the first asparagus and salad are ready for the table.[21] In William Cowper's
metaphrase of this popular image,

> The spiry myrtle with unwith'ring leaf
> Shines there and flourishes. The golden boast
> Of Portugal and western India there,
> The ruddier orange and the paler lime
> Peep through their polish'd foliage at the storm,
> And seem to smile at what they need not fear.
> Th' amomum there with intermingling flow'rs
> And cherries hangs her twigs. Geranium boasts
> Her crimson honors, and the spangled beau
> Ficoides, glitters bright the winter long.
> All plants of ev'ry leaf that can endure
> The winter's frown if screen'd from his shrewd bite,
> Live there and prosper. Those Ausonia claims,
> Levantine regions these; th' Azores send
> Their jessamine, her jessamine remote
> Caffraia; foreigners from many lands
> They form one social shade, as if convened
> By magic summons of th' Orphean lyre.[22]

The greenhouse redeems time, preserving summer throughout winter's darkest
months. In the kitchen garden, therefore, time is subject to artful design and
mapped onto carefully functional forms, the end result of which is profuse beauty:
sublime profligacy. Cowper's scene, like Switzer's narrative of the progress of the
peach from Persia to England, leads inexorably to the parable that, deftly trans-

planted from their own fiery lands, "foreigners" may be absorbed safely into England's benign shades.

Although the kitchen garden was supplied with its own historic antecedent in the ancient English messuage, garden manuals before the eighteenth century do not draw on this English heritage. Rather, authors locate gardening precedent in Eden, as Bacon's too frequently quoted beginning to "Of Gardening" suggests. Versions of his opening sally, "God Almighty first planted a garden," provide the starting point for many manuals and become a voluntary for garden histories. Later, however, even when authors use Bacon as their starting point, emphasis shifts from the idea that gardening is innocent, healthful, profitable, and delightful because of its Edenic antecedent to proofs that the English are particularly adept at the art.[23] Thus, in the first release of his *Gardeners Dictionary* (1724), Philip Miller could admit that "It is true the *Art* never arrived to any considerable *Pitch* in *England*, till within about thirty Years last past," but he goes on to extol the advances made by its patrons:

> But of late years many *Persons* of Fortune and Ingenuity have bent their Genius to the Study of it, and by that *Means* have not only set the professed *Gardeners* an Example, but also have generously given Encouragement to *Artists* to labour to trace Nature more closely in the *Propagation* of *Vegetables*, either *Trees, Plants, Flowers, Fruits*, &c. both for *Profit* and *Pleasure*, for *Use* and *Ornament*; so that of late Years, to their Praise be it spoken, it has been highly improved and brought to a considerable *Pitch*.[24]

There is an incipient sense of the national importance of the horticultural endeavor in this statement, particularly since Miller proceeds to anoint "Persons of Fortune and Ingenuity" as "*Patriots* of horticulture"—a clear rejoinder to Evelyn's canonization of "Paradisean and Hortulan saints" the century before.[25] Miller's 1724 preface, however, does not address a uniquely English tradition. As the century progressed, national endeavor was linked to English history in order to emphasize the innate productivity of the *English* kitchen garden and its distinctive attributes. In the process, the link to the biblical narrative of Eden became important primarily because England's soil offered a redemptive site in which the effects of the Fall could be reversed. By 1829 William Cobbett could retort, "I will not, with Lord Bacon, praise pursuits like these, because 'God Almighty first planted a garden;' nor with Cowley, because 'a Garden is like Heaven;' nor with Addison, because a 'Garden was the habitation of our first parents before their fall.'"[26] In the intervening years there grew a conviction, somewhat inflated, that kitchen and fruit gardens brought values of England's ancient past into the present. The messuage, a modern version of which would be nostalgically depicted in John Constable's painting of his father's kitchen garden in 1815, fed into a conservative aesthetic in which private property, independence, and the productive leisure of ordinary people—the frugal, hardworking, descendents of Saxon yeomen—were given an indigenously English cast.

The Thematics of Sexuality

The formal requirement of artifice central to the aesthetic of contained space caused theoretical difficulty for literary critics in the eighteenth century because a key reason for the deep suspicion of minor lyric was the continuity it maintained between artifice and sexuality. Following in the tradition of Petrarch and the ancients, the province of lyric was defined as love and seduction. In addition, the rich imagery of lyric poetry provided a precedent for linking verbal sensuousness with sensuality, so that the one, etymologically linked to the senses, achieves intelligibility as a result of its close homonymic and semantic relationship with the other, linked to eros.[27] As appropriations of phraseology from lyric verse prove, these discourses present a familiar body of literature that provided manuals on the enclosed, luxurious space of the kitchen garden with a strong sexual sensibility. At the same time, the confluence of the two bodies of literature succeeds in making sexuality in the garden appear to be inherent. This occurs in part simply because, as in lyric poetry, the *sensuous* language of garden manuals, which heaps the produce of the soil in a confined space, produces a distinctly *sensual* import. Deploring the adoption by essayists of lyric compression and disjunction that produced this effect, Dr. Johnson writes,

> A writer of later times has, by the vivacity of his essays, reconciled mankind to the same licentiousness in short dissertations; and he therefore who wants skill to form a plan, or diligence to pursue it, needs only entitle his performance an essay, to acquire the right of heaping together the collections of half his life, without order, coherence, or propriety.[28]

Although the "licentiousness" to which Johnson refers is more strictly formal than moral, it clearly implies that incontinence in compositional technique is homologous with sexual incontinence.

Despite Milton's powerful reinvention of the sonnet along political and meditative themes, the more conventional association between minor lyric and seduction remained strong during the eighteenth century. Thus, while shorter forms may have been thought more congenial to female intellectual capacities than georgic, the great ode, or epic, such subject matter could hardly recommend lyric poetry in an age when the definition of femininity was being moored to a maternal, nominally nonsexual image. The elegiac and domestic focus of Charlotte Smith's poems, first published in 1784, thus contested the sonnet's traditional subject matter and allowed her audience to imagine different themes for the

form—especially the social expectation that educated women reproduce normative family relationships. Smith helped to turn the sonnet from a Petrarchan
semantics of unrequitable desire toward the more bitter loss of the paternal home.
Second only to the loss of her father is that of her daughter through childbirth. The
normative structure of desire invoked by her sonnets is thus familial and reproductive rather than erotic.

Notwithstanding the thematic reorientation that Smith's sonnets afforded,
however, confined space remained freighted with erotic associations. Epic
episodes, as in Spenser's "Bower of Bliss," and conventional lyric contents form
part of a discursive tradition in which contained spaces are sexualized and
feminized—a tradition that was cemented by Milton's erotic depiction of both
Eden and Eve in *Paradise Lost*. This idea obtained across the literary spectrum. As
Johnson feared, it was not limited to minor lyric forms but spilled over into ex-
pository forms such as the essay and even the more empirical diction of the garden
manual. The sexualization of this horticultural site intensified through the eighteenth century, in part due to a strong tradition linking enclosed sites to female
sexuality that provided a cultural memory for such an identification. The increasing tendency over the course of the eighteenth century to revalue women's
sexuality as fundamentally maternal rather than erotic may seem alien to this
impulse, yet the two are allied. The wholesale adoption of Linnaeus's sexual system of botanical classification in the late 1750s made requisite an increasingly
specific sexual language in garden manuals; congruent with social expectations for
women, this sexual language focused attention on the reproductive function of
plants and assisted in the clarification of sexual behavior while naturalizing distinctions between the sexes.[29] Though the Linnaean system may seem an
adventitious factor in the history of the eros of contained space, this is not the
case. In a century dazzled by classification schemes, Linnaeus's botanical taxonomy was one of several available systems. Botanists and horticulturists adopted it
over other possibilities primarily, they argued, because of the simplicity of the
method. However, this should not obscure the fact that the sexual system was successful not merely because it was simple to apply, but because it seemed
commonsensical, feeding into an already existing notion that garden space was
erotic while confirming a developing logic of human sexual difference that focused
on maternity and productivity.

In theory, for a botanical sexual taxonomy to function scientifically, it should
filter out aspects of sexuality, such as desire, that are not reproductive. But while
Linnaeus made some attempt to isolate reproduction from other aspects of sexuality, he could not resist analogies to human sexual behavior that brought desire
within the compass of the scientific discussion of reproduction.[30] Thus, even
though the "sexual" classification system is manifestly about reproduction, it was
impossible to isolate reproduction in plants as a discrete and purely empirical

aspect of sexuality. As the connotative language of garden manuals demonstrates, sexuality cannot be contained by the limiting concept of reproduction, but is inevitably caught in the larger realm of the erotic. Each instruction for how to produce fruits and vegetables in such a system has the potential to bring with it other meanings. This penchant was abetted by the lyrical tendency in garden manuals to describe plants as active: twining, embracing, creeping, springing, catching hold of, clasping. So analogies between people and plants that energized the erotic quality of the garden supported emerging notions of sexual relations among people, providing visual images that quietly reified the continuities between sex, gender, and desire.[31]

The discourse of sexuality in the botanical realm thus converged in the eighteenth century with other discursive practices that mutually supported each other and rationalized public understanding of the kitchen garden, manuals for which promoted the new classification method while paving the way for the representational status of the contained space. The sexual tradition in plant classification handed down from earlier centuries, the science of human reproduction brought to the fore the previous century by William Harvey, and the erotic impulse in lyrical verse, which authors of garden manuals delight in quoting, together provided an analogical framework within which reproduction could be understood. Even economic assumptions gendered labor in the garden, thus contributing to the sexual framework of reproduction. Richard Weston's treatises, geared toward trade, illustrate the material gendering of garden labor. He advocated that in order to secure maximum profits from the family garden, women and their daughters were ideally suited as nurturers of silkworms, while children (with whom women were typically classified) could tend the flowers grown for the urban market.[32]

Myths also contributed. Samuel Collins concluded his manual, *Paradise Retriev'd* (1717), with the prurient claim that for the protection of melon beds,

> Ladies should not be invited to this Place, lest Nature should at that time prove in it's [*sic*] Venereall discharge, which has not only an Imaginary, but so real an Influence on Mellons newly set, that they will most of them drop off: I have found the consequence of this so fatal, that for many years last past (tho' they have been welcome to walk the rest of the Garden) I have been oblig'd at that time of fruiting, to deny their entrance into the Mellonry.[33]

Collins's superstition may seem fantastic, but it had an ancient precedent in garden literature. Thomas Hill advertised the idea in his discussion of cucumbers and gourds in 1563 and embroidered the notion over successive editions of his manual, citing as his authorities the Roman treatise writers, Columella and Florentius.[34] Conventional assumptions concerning the inherent eros of near-Eastern fruits gave the idea validity, as Marvell's familiar poem, "The Garden," illustrates. The sensual fifth verse reads:

What wond'rous Life in this I lead!
Ripe Apples drop about my head;
The Luscious Clusters of the Vine
Upon my Mouth do crush their Wine;
The Nectaren, and curious Peach,
Into my hands themselves do reach;
Stumbling on Melons, as I pass,
Insnar'd with Flow'rs, I fall on Grass.[35]

In Marvell's version of this idea the apple, like more erotic Persian fruits, is willingly fallen. Thus, even though Collins's conjecture was at worst rejected as an outright piece of villainy and at best quizzically received, response to it succeeded in circulating a titillating and familiar notion.[36] In recounting such a superstition, regardless of motive, gardeners and botanists open a window on a tradition in which ties between the fertility of the kitchen garden and sexuality by way of the female body were mythicized. Linnaeus's sexual system of classification reinforced the inherent sexuality in the garden that these discourses foster.

Linnaeus's sexual system achieved popular currency in England in part through Erasmus Darwin's *Loves of the Plants*, published in 1789. This four-book, quasi-georgic poem exemplifies the sexual assumptions I have outlined, working out the logical conclusions of the Linnaean system as a presentation of gendered relationships. Though the poem does not truly illustrate "the lascivious delight of a male-centered sexual fantasy," as has been charged, it mischievously exposes the sexual analogies between humans and plants that are such an important part of Linnaeus's premise.[37] Thus, tongue-in-cheek, Darwin succeeds in making the continuities between sex, gender, and desire intelligible. His humanized accounts of gendered relationships in the plant world, even when they may seem perverse according to Western cultural practice, all support a sexual system in which the male is active and dominant, the female is passive and receptive, domesticity is central, and heterosexuality is relentless. This is reinforced by the fact that the regulative structure of the system that Darwin's poem exposes is homosocial: all plants are defined by the male stamen if possible, especially in its relation to other stamens. The pistils (female) are defined only in relation to the male reproductive components. Contrary to objections of some of his critics, Darwin also maintains cultural codes for chaste behavior, as the titillating prospect of a description for Cerea, a flower characterized by twenty male components and one female, proves. In the face of obvious and more stimulating possibilities, Cerea is described as a virtuous damsel surrounded by lovesick but well-behaved suitors:

In crowds around thee gaze the admiring swains,
And guard in silence the enchanted plains;

> Drop the still tear, or breathe the impassion'd sigh,
> And drink inebriate rapture from thine eye.[38]

Other flowers with arrangements of stamens and pistils posing equally inventive possibilities are turned into sisters and brothers or, as in the case of Cyclamen, a maternal figure watching over her brood of sons. Nevertheless, it is significant that consistent with Linnaeus's system, the poem begins with Canna (*Monandria monogynia*), one male, one female. "The virtuous Pair" (1.41) thus forms the implicit standard against which all other flowers can be construed as variants. Somewhat perplexingly, and with droll results, Darwin never explains how generation takes place among his masses of siblings, virtuous maids, and well-behaved suitors.

Anna Seward defended Darwin's compositions, at least in part, as a matter of local patriotism; nevertheless, her points regarding the sexual content of the poem are well taken. In the cause of both his *Temple of Nature* and *The Loves of the Plants*, she protests,

> Young women who could be endangered by such descriptions, must have a temperament so unfortunately combustible as to render it unsafe to trust them with the writings of our best poets, whenever love is the theme. Paradise Lost presents more highly-coloured scenes than any which pass in the floral harems; so does the Song of Solomon, in which the language and images are infinitely more luxurious than the muse of botany ever exhibits.[39]

Seward's remark on the erotic language of *Paradise Lost* and *Song of Solomon* is shrewd. More to the point, however, both of her examples reinforce cultural sexual norms. Darwin's *Loves of the Plants*, in which even sexually anomalous orders such as the cryptogams are ingeniously brought under the standard of compulsory heterosexuality and appropriate female behavior, is equally assiduous. As Darwin amply reveals, Linnaeus provided a unified system in which sex is oppositional and desire is natural because it arises out of that opposition. The gender roles that were being so closely delineated throughout the last four decades of the century were thus overwhelmingly confirmed in his poem.[40]

These gender roles were confirmed by evolving descriptions in kitchen garden manuals for methods of growing cucumbers, a salad vegetable that had been an object of horticultural diversification since at least the sixteenth century despite the ill repute of its cool, damp humor. Descriptions of cucumbers and instructions for growing them reveal the transformation of botanical classification from morphology to reproduction, which provided a means of classifying plants by clearly delineated sexual functions. In the 1720s and early 1730s, prior to the introduction of the Linnaean system, Richard Bradley, Regius Professor of Botany at Oxford, was the prime English apologist for a theory of plant sexuality. Fruits and vegetables, in his view, were sexual organisms, a thought that he stoutly and falsely claimed "is entirely new, and seems reasonable."[41] Bradley's treatises are also an example of the way empirical language gives license to erotic diction. Thus, he claims concerning the function of bees in fertilizing flowers,

> that the *Farina Foecundus* or *Male Dust* has a Magnetick Virtue, is evident; for it is that only which they gather and lodge in the Cavities of their hind Legs to make their *Wax* with; and it is well known, that Wax, when it is warm, will attract to it any light Body. But again, if the Particles of this *Powder* should be required by Nature to pass into the Ovaries of the Plant, and even into the several *Eggs* or *Seeds* there contain'd, we may easily perceive, if we split the *Pistillum* of a *Flower*, that Nature has provided a sufficient Passage for it into the *Uterus*.[42]

Although this description may seem almost clinical to a modern reader, Bradley's diction was ridiculed by contemporary gardeners as overly sensual and anthropomorphized.[43] Bradley's sexual theory could be espoused, however, without an offensive sensualizing of the process of plant reproduction, as Philip Miller's descriptions of cucumbers demonstrated in the early editions of his *Gardeners Dictionary*. The restrained language of this gardener who was viewed as the foremost horticultural expert of his day provided a corrective to Bradley's human analogies. Though adopting Bradley's theory of plant generation, Miller ignored his erotic implications in the first edition of *The Gardeners Dictionary* (1724), simply distinguishing the flowers as male or barren and female or fruiting. However, successive editions, between 1731 and the posthumous edition of 1807, follow a significant pattern.[44] First, Miller adds "Characters"—precise botanical descriptions—to each entry. The Character is set off as a separate, italicized paragraph at the beginning of the entry, thus clearly distinguishing it from instructions for growing and cultivating the plants. In the first folio editions the Characters of cucumbers are mildly sexual. For example, in the edition of 1731 under *Cucumis* we find:

> It hath a Flower consisting of one single leaf, which is Bell-shap'd, and expanded towards the Top, and cut into many Segments, of which some are Male or Barren, having no Embryo, but only a large Style in the Middle, which is charg'd with the Farina; others are Female or Fruitful, being fasten'd to an Embryo, which is afterwards changed into a fleshy Fruit for the most part oblong, and turbinated, which is divided into three or four Cells inclosing many oblong Seeds.

By the seventh edition of 1759, however, when Miller reluctantly gave in to pressure to adopt Linnaeus's sexual taxonomy, he had tightened the Characters into an increasingly restrained and scientific diction. In the four lavishly illustrated folio volumes of the final, posthumous edition, the Character of the cucumber has been brought completely under the rule of science. Its sexual properties have been concealed in a diction that succeeds in providing full authority for the sexuality of plants while removing all traces of an erotic and lyrical language. The description, too long to reproduce in its entirety, is now organized as a series of clipped phrases, replete with abbreviations and Latin terms. The section on male blossoms alone reads:

> *Male flowers.
>
> Cal. Perianth one-leafed, bell-shaped, the margin terminated by five subulate teeth.
>
> Cor. five-parted, growing to the calyx, bell-shaped: divisions ovate, veiny-wrinkled.
>
> Stam. Filaments three, very short, inserted into the calyx, converging, of which two are bifid at the tip. The anthers are lines creeping upwards and downwards, outwardly adnate. Receptacle three-cornered, truncated, in the centre of the flower.[45]

The instructions for tending cucumbers in successive editions, between 1724 and 1807, follow an inverse pattern. Beginning in 1724, Miller, by comparison to Bradley, had allowed barely a hint of sexual language. With his adoption of the Linnaean system in the seventh edition, however, he overtly eroticizes the instructions:

> When the Fruit appears upon the Plants, there will also appear many male Flowers on different Parts of the Plant, these may at first Sight be distinguished; for the female Flowers have the young Fruit situated under the Flowers, but the male have none, but these have three Stamina in their Center with their Summits which are loaded with a golden Powder, this is designed to impregnate the female Flowers; and when the Plants are fully exposed to the open Air, the soft Breezes of Wind convey this Farina or male Powder from the male to the female Flowers, but in the Frames, where the Air is frequently too much excluded at this Season, the Fruit often drops off for Want of it: And I have often observed that Bees that have crept into the Frames when the Glasses have been raised to admit the Air, have supplied the Want of those gentle Breezes of Wind; by carrying the Farina of the male Flowers on their hind Legs into the female Flowers, where a sufficient Quantity of it has been left to impregnate them. For as the Bees make their Wax of the Farina or male Powder of Flowers, so they search all the Flowers indifferently to find it; and I have observed them come out of some Flowers with their hind Legs loaded with it, and going immediately into other Flowers which have none, they have scattered a sufficient Quantity of this Farina about the Style of the female Flowers to impregnate and render them prolifick. These Insects have taught the Gardeners a Method to supply the Want of free Air, which is so necessary for the Performance of this in the natural Way; this is done by carefully gathering the male Flowers, at the Time when this Farina is fully formed, and carrying them to the female Flowers, turning them down over them, and with the Nail of one Finger gently striking the Outside of the male, so as to cause the Powder on the Summits to scatter into the female Flowers, and this is found to be sufficient to impregnate them, so that by practising this Method, the Gardeners

have now arrived at a much greater Certainty than formerly to procure an early Crop of Cucumbers and Melons, and by this Method the new Varieties of Flowers from Seeds, which is done by the mixing of the Farina of different Flowers into each other.[46]

In this edition, Miller's bees have become promiscuous fellows, creeping into the cucumber's greenhouse home with their tribute carried on their hind legs to perform the fructifying office when breezes are denied. Furthermore, rather than merely perform their office, they instruct the gardener in his art. Augmenting the erotic content, the gardener's finger becomes the means by which the male flower's potency can be restored. The lyricism of the instructions has also deepened. Golden powder, soft breezes, and scattering pollen provide rich mental images, while activity itself has been theatricalized. The breezes, the bees, and the gardener all play a part in a delightful performance, the end result of which is fruition. Notably, Miller repeats Bradley's conclusion, now fortified by Linnaeus's work, that it is thus that the astonishing variety of vegetables or fruits within a given kind has been produced.

In this context, Cowper's metaphrase of Miller's instructions in the third book of *The Task* is instructive. Cowper retains Miller's instructions while sanitizing them of their erotic content. His version, while retaining the vivid, active language of kitchen garden manuals, is nevertheless chaste. On the cultivation of his beloved cucumbers he merely comments,

> These have their sexes, and when summer shines
> The bee transports the fertilizing meal
> From flow'r to flow'r, and ev'n the breathing air
> Wafts the rich prize to its appointed use.
> Not so when winter scowls. Assistant art
> Then acts in nature's office, brings to pass
> The glad espousals and insures the crop. (*The Task* 3.537–43)

Cowper's reluctance to displace the sexual into the erotic suggests the options that authors of kitchen garden manuals were free to choose or reject. Indeed, in contrast to Cowper, John Abercrombie elected to increase the erotic effect of Miller's instructions. Taking full advantage of the gardener's role in assisting the cucumber to fruition, he advises,

> Likewise, when the plants are in blossom take care to impregnate the female flower with the *anthera* of the male, which you will pluck for that purpose the same day that both flowers first expand; pulling away the flower, leaf or petal of the male blossom, then holding the shank betwixt the finger and thumb, introduce the anthera into the center of the female flower, touching the stigmata thereof, twirl it about so as to leave some of the male powder upon the female organ, then throw it away, this completes the business, renders the flowers fertile, and the fruit sets freely.[47]

Significantly, by 1807 the role of the gardener in reproducing cucumbers would be qualified. The editor of the posthumous edition of Miller's *Gardeners Dictionary*, Thomas Martyn, notes that "it is probably less necessary to carry the males to [the female flowers], as practised by some gardeners, though nature having provided male flowers, it is most likely that the pollen in the anthers of the others is frequently defective."[48] Despite his reservations, however, he leaves intact the lyric residue of Miller's instructions, thus indicating the power of convention to define practice even in the face of new data: soft breezes still carry the golden, inseminating powder to receptive female flowers; in their absence, the promiscuous bee fulfills his potentizing office; and, more reliably, despite Martyn's expressed reservations, the gardener supplies what the greenhouse wants, bringing the male flower to the female and facilitating the act of impregnation with a gentle tap of his finger. The telos of the reproductive system is revealed in the cucumber's prolific varieties, which, we are told, now include the "gourd, globe cucumber, round prickly-fruited cucumber, African cucumber, acute-angled cucumber, common or musk melon, apple-shaped cucumber, hairy cucumber, common cucumber, serpent cucumber or melon, flexuose cucumber or melon."[49] The category of common cucumber alone may be distinguished by eight different subspecies.

The inverse relationship between the increasing restraint of the *Dictionary*'s Characters and intensifying sensual and lyric language in the instructions of successive editions has a subtle logic. The Linnaean taxonomy, which systematized and regulated botany over the latter half of the eighteenth century, brought scientific legitimation to the sexual organization of plants. Darwin's theatricalization of plant sexuality and Linnaeus's anthropomorphisms simultaneously sanctioned broad parallels between plant and human sexuality. Thus, botany achieved explanatory power in the realm of human behavior, even as human behavior was used to illuminate the secret life of plants. Overt sexual parallels, however, were eventually erased. A neutral, elisionary, scientific language naturalized these parallels even as it cloaked the sexual and erotic vocabulary that had given them power in the first place. Botany thus helped to legitimate emerging sexual distinctions as a part of the nature of things, not least since it was used as a tool in the education of girls. In the pursuit of taxonomy, notions of sexuality that formed the core of this system were absorbed along with the scientific language that sanitized them. The sexualization and desexualization of the cucumber in the *Gardeners Dictionary* contributed to the public logic in which women in the eighteenth century were transformed from problematically sexual beings, whose reproductive capacities were a part of their erotic capacities, into maternal beings. The regulatory principle of the kitchen garden that plants be brought to fruit epitomized the transformation in the cultural realm whereby women were brought under the rule of maternity, with its suppressed erotic connotations. The kitchen garden thus provides a component of a larger signifying system that was used to elaborate normative human distinctions.

Linnaeus's sexual classification of plants may seem to have taken us far from Abercrombie's "glassary compartments" teeming with vegetables, flowers, and fruits, and even further from the literary world of the lesser lyric. Yet Abercrombie's marvelously lyrical phrase epitomizes the enigmatic relationship between private sentiment and public domain that lyric verse presumes. Patricia Fumerton has analyzed the relationship between private and public in the Renaissance sonnet. She concludes, "However much we may need to define the concepts as separate (or envision a culture *all* one or the other), 'private' and 'public' can only be conceived as a split unity divided along a constantly resewn seam that can never be wholly closed or absolutely parted."[50] The discourse of the kitchen garden, which finds its distillation in the greenhouse, reveals just this tenuousness. Although the walls of the greenhouse are transparent for the modern scientific purpose of exposing plants to heat and sunlight, they also, in the most ancient purposes of the wall, command an exclusionary function. As the word *compartment* suggests, the greenhouse is a world set apart that functions according to its own laws. It connects to the discourses of lyric and the female body in this salient way, providing the illusion of openness when it is, in actuality, compartmentalized and guarded. Like the female body, which was enlightened as science revealed its reproductive depths, yet more guarded as it was set apart by the ideology of domesticity, the kitchen garden revealed its horticultural contents while presenting itself as the most guarded of topographical spaces. In the historically specific circumstances of the late eighteenth century, compartmentalization and guardedness became linked, naturally it would seem, to productivity. The discourses of the female body and the kitchen garden that I have documented here thus epitomize the problem of lyric verse with its riddle of private sentiment that nevertheless assumes productivity in that it demands to be published. As this riddle affirms, despite the social and political attractions of the unbounded view that governed the aesthetic of landscapes and poetic forms for the majority of the eighteenth century, shifts in public perception made requisite a realignment between productivity and enclosure along a continuum of aesthetic, functional, and sexual forms.

Notes

1. Susan Stewart, *On Longing: Narratives of the Miniature, the Gigantic, the Souvenir, the Collection* (Durham: Duke UP, 1993), 6.

2. See Thomas Laqueur, *Making Sex: Body and Gender from the Greeks to Freud* (Cambridge: Harvard UP, 1990), especially chs. 5 and 6.

3. Henry Home, Lord Kames, *Elements of Criticism, The Third Edition. With Additions and Improvements*, 2 vols. (Edinburgh: A. Millar, A. Kincaid, J. Bell, 1765), 187.

4. See Richard Weston, *Tracts on Practical Agriculture and Gardening* (London: S. Hooper, 1769), 42–43; and Humphry Repton's Redbook for Blenden Hall excerpted in *Fragments on the Theory and Practice of Landscape Gardening* (London: J. Taylor, 1816), 23–28.

5. W. R. Johnson, *The Idea of Lyric: Lyric Modes in Ancient and Modern Poetry* (Berkeley and Los Angeles: U of California P, 1982), 82.

6. See Richard Bradley, *A General Treatise of Husbandry and Gardening*, 2 vols. (London: T. Woodward and J. Peele, 1726), 2:246; and George Mason, *An Essay on Design in Gardening* (Dublin: John Exshaw, 1770), 25–26.

7. Batty Langley, *New Principles of Gardening* (London: A. Bettesworth et a., 1728), 193.

8. Philip Miller, *The Gardeners and Florists Dictionary, or a Complete System of Horticulture, &c.*, 2nd ed. (London: Charles Rivington, 1731; 1st ed. 1724), v; hereafter cited by year of edition.

9. Repton, *Fragments*, 27–28.

10. Francis Bacon, "Of Gardens," *The Works of Francis Bacon*, ed. James Spedding, Robert Leslie Ellis, and Douglas Denon Heath, 12 vols. (Boston: Brown and Taggard, 1860), 12:235; Miller, *Dictionary* (1724), vii.

11. Stephen Switzer, *The Practical Kitchen Gardiner* (London: Thomas Woodward, 1727), 396–97; Jean de la Quintinye, *The Compleat Gard'ner*, trans. John Evelyn (London: Matthew Gillyflower and James Partridge, 1693), 6:179.

12. Richard Weston, *The Universal Botanist and Nurseryman*, 2nd ed., 4 vols. (London: J. Bell, 1777), 1:iv.

13. *The Practical Fruit-Gardener* (London: Thomas Woodward, 1724), 3–4.

14. Ibid., 81.

15. Ibid., 78–80.

16. *Paradise Lost, Complete Poems and Major Prose*, ed. Merritt Y. Hughes (Indianapolis: Odyssey Press, 1957), 4.206–08; hereafter cited parenthetically in the text as *Paradise Lost* with book and line number(s).

17. Switzer, *Kitchen Gardiner*, xx. He quotes from Evelyn's translation of Quintinye's calendar entry for April (6:179).

18. *The Hot-House Gardener* (London: John Stockdale, 1789), 127–28.

19. Weston, *Tracts on Practical Agriculture and Gardening* 194–95.

20. *The English Garden: A Poem*, 4 books (London: J. Dodsley et al., 1778–81), 2.21–22.

21. The term *glassary apartments* is Abercrombie's (*Hot-House Gardener*, 123, 128, 185).

22. *The Task, a Poem in Six Books*, in *The Poems of William Cowper*, ed. John D. Baird and Charles Ryskamp, 2 vols. (Oxford: Clarendon, 1995), 3.570–87; hereafter cited by book and line number in the text.

23. George Johnson, *A History of English Gardening Chronological, Biographical, Literary, and Critical* (London: Baldwin and Cradock et al., 1829), 1.

24. *Dictionary* (1724), 9.

25. *Dictionary* (1724), xii; John Evelyn, "Garden Letters," *Sir William Temple upon the Gardens of Epicurus, with Other XVIIth Century Garden Essays*, ed. A. F. Sieveking (London: Chatto and Windus, 1908), 176.

26. *The English Gardener* (London: n.p., 1829), n. pag., par. 58.

27. As the *Oxford English Dictionary* indicates, Milton invented the distinction between the words *sensuous* and *sensual*, but it was not until Coleridge put the word *sensuous* into use that the distinction was commonly made. See *The Notebooks of Samuel Taylor Coleridge*, ed. Kathleen Coburn, Bollingen Series 50, 4 vols. to date (Princeton: Princeton UP, 1961), 2:2442. Although this precise vocabulary may not have been common in the intervening century, comments of literary critics, especially Samuel Johnson, indicate the erotic associations of richly sensuous lyric diction.

28. *The Rambler*, in *The Yale Edition of the Works of Samuel Johnson*, eds. W. J. Bate and Albrecht B. Strauss, 16 vols. to date (New Haven: Yale UP, 1969), 3:77.

29. Ann B. Shteir provides an important analysis of the cultural contexts and effects of Linnaeus's classification system in her study of women in botany: *Cultivating Women, Cultivating Science: Flora's Daughters and Botany in England, 1760 to 1860* (Baltimore: Johns Hopkins UP, 1996), 9–32.

30. John Farley quotes a suggestive example. In an extended metaphor, Linnaeus imagines that "The *calyx* then is the marriage bed, the *corolla* the curtains, the filaments the spermatic vessels, the *antherae* the testicles, the dust the male sperm, the *stigma* the extremity of the female organ, the *style* the *vagina*, the *germen* the ovary, the *pericardium* the ovary impregnated, the seeds the *ovula* or eggs." *Gametes and Spores: Ideas about Sexual Reproduction, 1750–1914* (Baltimore: Johns Hopkins UP, 1982), 7.

31. Throughout this essay, my understanding of the continuities between sex, gender, and desire and the notion of intelligible gender is drawn from Judith Butler, *Gender Trouble: Feminism and the Subversion of Identity* (New York: Routledge, 1990), see especially 16–25.

32. Weston, *Tracts*, 28, 62.

33. *Paradise Retriev'd* (London: John Collins, 1717), 106. Collins was not considered simply a crackpot: his methods for raising cucumbers served as a source for Philip Miller's *Dictionary*.

34. *First Garden Book: Being a Faithful Reprint of A Most Briefe and Pleasaunt Treatvse, Teaching howe to Dress, Sowe, and Set a Garden*, ed. Violet and Hal W. Trovillion (1563; Herrin, IL: Trovillion Private Press, 1939), n. pag. The superstition is repeated in editions of 1568, titled *The Proffitable Arte of Gardening* (178v.), and 1608, titled *The Arte of Gardening* (151).

35. *The Poems and Letters of Andrew Marvell*, ed. H. M. Margoliouth, 3rd ed., rev. Pierre Legouis and E. E. Duncan-Jones, 2 vols. (Oxford: Clarendon, 1971), 1:52; lines 33–40.

36. An anonymous pamphlet published shortly after Collins's singles out this theory as indecent; see *A Letter from a Gentleman in the Country to his Friend in Town: concerning two Books lately published by Mr. Bradley, and Mr. Collins* (London: Bernard Lintot, 1717), 33–34. Bradley, while quizzical, does not deny the notion's credibility; see *New Improvements of Planting and Gardening, both Philosophical and Practical*, 6th ed. (London: J. and J. Knapton, 1731), 265. Informal discussions I have had with a variety of people indicate that the superstition is still alive.

37. Shteir, *Cultivating Women*, 27.

38. *The Loves of the Plants*, in *The Botanic Garden, A Poem. In Two Parts. Part I. containing The Economy of Vegetation. Part II. The Loves of the Plants. With Philosophical Notes*, 4th ed., 2 vols. (London: J. Johnson, 1799), 4.29–32; hereafter cited in the text.

39. *Letters of Anna Seward: Written between the years 1784 and 1807*, 6 vols. (Edinburgh: Archibald Constable et al., 1811), 6:142.

40. As Darwin points out in his treatise on education for girls, "I forbear to mention the Botanic garden; as some ladies have intimated to me, that the Loves of the Plants are described in too glowing colours; but as the descriptions are in general of female forms in graceful attitudes, the objection is less forceable in respect to female readers." *A Plan for the Conduct of Female Education in Boarding Schools* (Derby: J. Johnson, 1797), 38.

41. *Husbandry and Gardening*, 1:333. Although Bradley's notion of plant sexuality was not new, he does seem to have formulated a theory that both pollen and seed contribute characteristics to the new plant well before Linnaeus's work became known in England.

42. *Planting and Gardening*, 11.

43. Both Switzer and Langley reject Bradley's insinuations. Switzer, *Kitchen Gardiner*, 112; Langley, *New Principles of Gardening*, 38.

44. Miller's instructions for planting melons are from the beginning more sexualized than those of cucumbers, despite the fact that the characters are very similar. In the case of melons, Miller describes the importance of "impregnat[ing] the Ovary of the fruitful Flowers: Which when done, the Fruit will soon swell and grow large." Completing the image, he claims that if his instructions are heeded, "there is no Danger of miscarrying." *Dictionary* (1731), "Melon," arranged alphabetically. This is not surprising given the erotic association of melons as exotic oriental fruits in the lyric tradition. In later editions of the *Dictionary*, Miller's instructions for tending melons replicate those for tending cucumbers.

45. *The Gardener's and Botanist's Dictionary*, ed. Thomas Martyn, 4 vols. (London: F. C. and J. Rivington et al., 1807), alphabetical.

46. Miller's description amplifies the method for fertilizing flowers described by Linnaeus in a prize-winning lecture at the Imperial Academy at Petersburg in 1760. Carolus Linnaeus, *A Dissertation on the Sexes of Plants*, trans. James Edward Smith (London: George Nicol, 1786), 36–39.

47. John Abercrombie, *The Complete Forcing-Gardener* (London: Lockyer Davis, 1781), 66–67.

48. Miller, *Dictionary* (1807), alphabetical.

49. Ibid.

50. *Cultural Aesthetics: Renaissance Literature and the Practice of Social Ornament* (Chicago: U of Chicago P, 1991), 109–10.

Chapter Eight

Romantic Exemplarity

Botany and "Material" Culture

Theresa M. Kelley

With what curious attention does a naturalist examine a singular plant, or a singular fossil, that is presented to him? He is at no loss to refer it to the general genus of plants or fossils; but this does not satisfy him and when he considers all the different tribes or species of either with which he has hitherto been acquainted, they all, he thinks, refuse to admit the new object among them. It stands alone in his imagination, and as it were detached from all the other species of that genus to which it belongs. . . . When he cannot do this, rather than it should stand quite by itself, he will enlarge the precincts, if I may say so, of some species in order to make room for it; or he will create a new species on purpose to receive it, and call it a Play of Nature, or give it some other appellation under which he arranges all the oddities that he knows not what else to do with. But to some class or other of known objects he must refer it, and betwixt it and them he must find out some resemblances or other, before he can get rid of that Wonder, that uncertainty and anxious curiosity excited by its singular appearance, and by its dissimilitude with all the objects he had hitherto observed.

—Adam Smith

In 1753, Sir Hans Sloane's last will and testament bequeathed to the soon-to-be-established British Museum his magnificent collections, together with his hope that they would be exhibited with the same degree of systematicity that had made his botanical collection so celebrated among his contemporaries.[1] For eighteenth-century writers on aesthetics, the alternative to orderly presentation was a disabling plenitude in the early exhibitions of the Museum and in nature itself, as though Smith's "Play of Nature" had managed to reproduce itself many times over. Smith's regulative anxiety is not unique. If confronted with an "infinity of objects," Francis Hutcheson advised in 1725, "heaping up a Multitude of particular incoherent Observations" makes for an "uneasy state of Mind" unless "we can discover some sort of Unity, or reduce them to some general Canon."[2] Hutcheson and Smith specify the problematic ground of my argument. Echoing down the corridors of the British Museum in the first decades of its existence, versions of Hutcheson's nervous preference for "unity in variety" sought to manage the burgeoning record of flora and fauna within taxonomies that were anchored by Linnaean systematics until the second decade of the nineteenth century.

Until the publication of Charles Darwin's *The Origin of Species* (1859), the English preoccupation with botanical taxonomy and nomenclature was as motivated by the quest for unity as it was by the effort to codify botanical variety. During the Romantic era (1780–1840), "new" or exotic flora and continued scrutiny of newly discovered aspects of known flora put continuous pressure on the desire to classify, reclassify, and name species and the systematic arrangements to which species were said to belong. I argue in this essay that Romantic debates about botany and nomenclature belong to, and may have helped to incite, a wider cultural and literary attention to classification and naming and, further, that this attention complexly informs how two Romantic poets—Charlotte Smith and John Clare— use poetic figures derived from botanical names.

The critical itinerary of this argument is problematic, and for that reason it is also instructive about the difficulties that ensue when critics gesture across aesthetics to culture and back again. First, it imagines or finds—the ambiguous doubling of verbs is suggested by their common Latin root *invenio*—an adequate conceptual framework for the role of botany in Romantic culture and, in particular, for its role in the poetry of Charlotte Smith and John Clare. As a cultural history, this claim joins two essentially dissimilar levels of history. On one level my argument alludes to evidence roughly comparable to what Siegfried Kracauer calls "micro history," which emphasizes archival (or textual) evidence pertaining to a specific place and time. On the level of macro history, where different levels of generality and abstraction pull away from the micro event or detail,[3] I argue that there is something exemplary for thinking about Romanticism in the way (or ways) Smith and Clare use botanical figures. Put this way, the problem opens in another direction. For inasmuch as the "micro" level of my argument concerns poetic figures that recognize or resist taxonomic distinctions and labels, it joins a record of individual poetic agencies to that larger cultural set.

Second, this essay is prompted by a critical desire that is on display in this volume and in recent scholarship on the cultures of Romanticism. At the very least, my argument about Romantic botany and poetic figures attempts or imagines joining Romanticism to its "material" cultures, to the "materiality" of culture or its "material ground" (a double figure that seeks to ensure its own ground without proving it on some other ground). On this point I am willing to risk some bravado and more tautology: I claim, in brief, that the Romantic attention to taxonomy, including poetic attention, registers in a surprising, because seemingly remote, corner of that era a trenchant instance of Romanticism's preoccupation with representation and representability—with the problem of relating parts to wholes or even finding wholes for those parts, whether the inquiry is aesthetics, politics, philosophy, or natural history.

I approach this topic via botany in part because it offered English writers an early and ready engine for thinking about taxonomy as a way to map variety. As the first branch of natural history to have engaged amateur English naturalists in the seventeenth century,[4] botany offered subsequent writers familiar and thus useful illustrations for various topics, including the problem of order and diversity. For example, the title and ostensible subject of Adam Smith's essay from which I have extracted the remarks that appear at the beginning of this chapter concern not natural history but astronomy. Smith's full title is "The Principles which lead and direct philosophical Enquiries; illustrated by the History of Astronomy."[5]

My second reason for considering natural history, specifically botany, is more explicitly polemical. As material cultures go, this one is as "material" as it gets. A study like this one is thus obliged to think through the strong desire to map or remap Romanticism by way of what have been characterized as its material cultures. I admit complicity in that desire. The critical romance in the title of Bill Brown's *Material Unconscious* is as much my romance as it is Romantic. In that study Brown claims that literature, specifically American literature, consciously and unconsciously "accretes and figures sociohistorical fragments" such that the outcome recasts "relations between the particular and the general, the material and the conceptual, the synchronic and the diachronic, the local and the national."[6] My argument about the usefulness of botany for Romantic culture and specifically for two Romantic poets is evidently on the lookout for those relations. And precisely for this reason, it is inherently less disposed to acknowledge gaps bridged or papered over by such claims. In acknowledging them here, I recognize that the argument of this essay repeats the gesture embedded in Smith's call for taxonomies to map an otherwise unmanageable set of curiosities. The search for Romantic exemplarity—for poetic parts to attach to the whole we call Romanticism—echoes, in other words, a familiar gesture in Romantic writing, one already specified before Samuel Taylor Coleridge made strong and reiterated claims for the organic relations among parts and wholes. If mine is a critical tautology, it is at least one with an impeccable Romantic lineage.

Before Sloane died, and more conspicuously afterward, English botany was, as indeed were all branches of natural history, absorbed by the task of classification and nomenclature. One provocation for taxonomic inquiry was the English acceptance of Linnaeus's binomial Latin nomenclature and system by the 1750s, which quickly replaced or at least marginalized the messy array of names and nicknames for plants that had been invented at different times and in different languages and dialects across the known world. Linnaeus's systematics explicitly addressed the discovery of new and strange species from around the globe that had already begun to put extraordinary pressure on the task of classifying its flora and fauna. The earnest temperament of taxonomic industry in England and on the Continent moved in restrained counterpoint to the promiscuous sexuality, plant mimicry, and teasing residuals of English plant names that disturbed taxonomic debates from Erasmus Darwin to Charles Darwin. Although this disturbance was not of the same order as the Victorian Darwin's theory of evolutionary adaptation and the emergence of new species, it nonetheless unnerved the strong taxonomic desire to identify immutable species.

The logic of this disturbance, to which my argument will return, was metaphoric. For by introducing hybrid, lawless sexual transactions among flowers and insects, Linnaeus's sexual system made possible the exuberant metaphoric language of Erasmus Darwin's *The Botanic Garden* (1791) and that of subsequent publications such as Robert Thornton's *Temple of Flora* (1798–1807) and, much later, James Bateman's *The Orchidaceae of Mexico and Guatemala* (1837–1843). The explanatory ground for my present argument, the role of botanical figures in poems by Charlotte Smith and John Clare, requires some notice of historical threads that this essay bundles into a sheaf: taxonomy, European exploration, and botanical discovery.

Throughout the century of scientific and popular presentations of species, variations, and competing systems for their classification that had begun in the decades before Sloane's death and ended, more or less, with Darwin's publication of *The Origin of Species*, taxonomy was hardly a model of Kantian disinterestedness, not even for Kant, whose remarkable meditations on aliens and hybrid species respond to concerns that preoccupy writers before and after Kant.[7] Rather, the work of taxonomy had interests that were global and local, mercantile, prayerful, and self-serving, by turns or all at once. By the late eighteenth century, botanical collection, preservation, and illustration had become strategic mechanisms in the monumental British and Continental effort to codify and thereby possess the unexpected botanical plenitude that accompanied the discovery of "new" worlds, or worlds new to the West. In a nicely ironic recapitulation of the Adamic relation between naming and property, European explorers who identified and named the new and thus exotic plants and animals also brought them home to grow in nurseries on "native" soil.

The spur to incorporate more kinds into various taxonomic systems is also apparent at home, where the diversity of native kinds is a domestic version of the proliferation, as it were, of exotic kinds. Those responsible for the accumulation of specimens and classificatory knowledge included well-known figures such as Sir Joseph Banks and Charles Darwin and many who were less well known or anonymous. In the latter category were women, dissenters, and "native" artists and informants across the globe who gathered, classified, and drew or engraved plants as well as birds, insects, and other artifacts of natural history. The products of all this activity were preserved in herbaria, nurseries, and gardens and were disseminated in botanical paintings and illustrations, manuscripts, journals, and books.[8]

Given so much natural and corresponding cultural plenitude under the banner of natural history, taxonomy offered a mechanism and capable fiction for control over a natural world whose proliferation of species—at least for Western eyes—elicited a desire to master alien kinds that paralleled the effort to map and acquire new lands for commerce, slavery, and plantation. At The Royal Botanic Gardens at Kew and in London, Banks became a "center of calculation" for natural history and particularly botany as informants from around the globe sent back specimens and information.[9] During the Romantic era, as more exotic species were introduced into English and European gardens, local and exotic species competed for space in the rage for botanical illustrations in magazines and books. The descriptions that accompanied these illustrations typically noted when the exotic plant arrived in England or Europe and when it first bloomed as a nativized species. The published record of this traffic from exotic to domesticated status mirrors the larger transactions that constitute the history of European imperial and colonial expansion.

Bringing exotic plants home to England to make them grow and bloom at home (the pineapple), or transporting English plants to distant colonies (English fruit trees to Australia) or from one distant colony or at least port of call to another (the breadfruit tree from the South Sea islands to Jamaica) were thus strategies for domesticating an exotic plenitude that was by turns strange and strangely familiar.[10] For example, the Dragon Arum (Figure 8.1), presented as a huge, ominous blossom in Thornton's *Temple of Flora* (1798–1807), has among its English familiars *Arum maculatum*, identified as Cuckow-pint or Wake Robin (Figure 8.2) in the collaborative, multivolume Sowerby/Smith *English Botany*, and so identified by both Charlotte Smith and John Clare.[11]

The history of taxonomy and nomenclature since the Renaissance is a Byzantine record of different systematics and assumptions, of slight modifications, radical innovation, and incremental or wholesale adjustments in taxonomic categories, hierarchies, and nomenclatures.[12] The proliferation of information about new and alien kinds or species fueled efforts to devise scientific taxonomies of plant and animal kinds and involved its practitioners in a teasing paradox. Although the subject of taxonomic inquiry is nature, the "natural histories" that emerged were anything but natural. Yet from Aristotle to the present, the philosophical conviction that there might be "essential kinds" in nature has complicated or at times obscured this paradox.

Figure 8.1. Robert Thornton, *The Temple of Flora* (London, 1799–1807), Plate XXII: Dragon Arum. Mezzotint engraving (aquatint in final state) by W. Ward from a painting by Peter Henderson. Courtesy of the Linda Hall Library, Kansas City, MO.

Figure 8.2. William Curtis, *Flora londinensis: or Plates and descriptions of such plants as grow wild in the environs of London,* 2 vols. (London: W. Curtis, 1777ff.), Vol. 1, Plate 114: *Arum maculatum* (Cuckow-pint or Wake Robin). Hand colored engraving by Sansom from a drawing by James Sowerby. Courtesy of the Linda Hall Library, Kansas City, MO.

The philosophical status of "natural kinds" and classificatory schemes for organizing them has turned—with varying degrees of commitment to and against
this view—on whether "kinds" or "species" or whatever name one chooses to give
them refer to essences to which all particulars or their most basic groupings could
be said to refer. Scott Atran contends that even in folk biologies, the preferred
taxonomic framework is one that includes all known kinds. This disposition exceeds the more limited taxonomic ambition to recognize only or primarily useful
kinds, the project which others have ascribed to folk cultures (i.e., this plant is poisonous, that root beneficial).[13]

Precisely because folk cultures typically sort for an inclusive taxonomic array
of the flora and fauna of a specific locale, those taxonomies cannot predict the identity and classificatory relation among natural kinds found elsewhere. Even when
they have occasion to recognize related kinds in a new environment, they do not
use those occasions to predict the presence of natural kinds outside a known environment. Approximating something akin to modern inquiries about ecosystems,
folk taxonomies are less interested in hierarchies and their causes—the founding
ambition of modern scientific systematics—than in a spontaneous, orderly group-
ing of life-forms in a locality.[14] In sharp contrast, the rise of scientific taxonomies
in the sixteenth and seventeenth centuries was in some measure prompted by the
need to predict what species might be found outside the known world, and thereby
to anticipate and place new species in existing hierarchies.

Although the seventeenth-century physician John Ray and subsequent tax-
onomists were attentive to morphological differences as the basis for taxonomic
arrangements of species, genera, and families across many domains, they disagreed
about how to hierarchize those differences. Ray published taxonomic theory and a
dictionary in Latin, but he included traditional English plant names from different
localities and insisted that local informants, whom he identified by name, provided
valuable intuitions about "true genera."[15] Ray tempered this epistemological and taxonomic hope along lines suggested by his contemporary John Locke, whose brief for
the limits of human knowledge surfaces in Ray's admission that the most likely access
to knowledge of genera lay in a flexible attention to morphological affinities.

The contrast between Ray's late-seventeenth-century preference for local informants, names, and intuitive grasp of possible genera and Linnaeus's
early-eighteenth-century introduction of a binomial Latin nomenclature and taxonomy is striking. For Linnaeus, accumulating evidence of new species demanded
a system, admittedly artificial, in which all plants and animals could be identified
by two names, rather than a string of descriptive names or traditional local names
that could and did vary from region to region. To construct this system, Linnaeus
argued that the reproductive parts of plants—their stamens and pistils—could be
classified to include all—well, mostly all—flowering kinds. The twenty-four classes
of the Linnaean system refer (in the first twelve) to the number of stamens, to the

length or groupings of stamens (classes 12–19), more complicated sexual arrangements, including those hermaphroditic flowers that are so variously engaged in Erasmus Darwin's *Loves of the Plants,* part 2 of *The Botanic Garden* (classes 20–23), and the one class that puzzled Linnaeus's taxonomic itinerary, the Cryptogamia, so called because members of this class lacked proper flowers and so reproduced themselves by rather cryptic (as opposed to sexually explicit) means.[16]

Couched in the invented Latin tags that could be easily memorized as long as one was persuaded by the explanatory logic of the system those tags marked, Linnaean systematics were widely accepted in England by 1760 as the method for learning about natural history that was accessible to all who would commit themselves to learning enough Latin to use it. Many did, including botanists of the working or middling classes who would not otherwise have studied Latin, such as John Bartram in Philadelphia and James Lee, the Quaker nurseryman from near London who, among many others, wrote an introduction to the Linnaean system and provided knowledge and specimens for botanical illustration in popular magazine and book formats, including elephant and double-elephant folio volumes that were expensive to produce and in their own way monumental.[17]

English writers and botanists defended the Linnaean system long after widespread recognition on the Continent that it could not be reconciled with evidence provided by the investigation of plant morphology and the discovery of new species. By the early 1820s, English botany had turned from Linnaean systematics to one based on the so-called Natural Orders. Beginning in 1789, Antoine Laurent de Jussieu, then Michel Adanson, and finally Augustin-Pyrame de Candolle successively refined this new system of taxonomy, which determined which plants belonged to which orders and families by scrutinizing all morphological affinities before deciding which would be chosen as leading indicators of taxonomic relationships. In the Linnaean system, the only morphology that counted for taxonomic purposes was that of the sexual, hence reproductive, parts of a flower.[18]

It is perhaps not surprising that the Cryptogamic plants, which lacked visible sexual parts, became poster plants for what was unsatisfying about the sexual system (Atran 180, 185–86, 198–99). Some flickers of disagreement with Linnaean classification occasionally occur in the brief descriptions of plant illustrations in magazines and books, which consistently noted when experts disagreed about a species or genus designation. A typical entry would indicate the Latin nomenclature for the plant, a common name (or several) followed by its taxonomic designation, and a list of works in which the plant is discussed. Additional comments might concern habitat and, if pertinent, an explanation of taxonomic disagreement. The *Arum maculatum* entry in the Sowerby/Smith *English Botany* is instructive:

ARUM maculatum. *Cuckow-pint,* or *Wake Robin*—MONOECIA Polyandra. . . .
Linn. Sp. Pl. 1370. Sm. Fl. Brit. 1024. Hudson 395. With. 497. Hull 197. Relh.

353. Sibh. 177. Abbot 197. Curt. Lond. Fasc. 2.t. 63. Woodv. Med. Bot.t.25. Arum. Raii Syn. 266. . . . This position of the flowers, exactly analogous to that of the most genuine monoecious plants, has induced us to remove this genus to the *Monoecia.* No principle can reconcile it to the *Gynandria.*[19]

As the editor and compiler of *English Botany,* Sir James Edward Smith, first president of the Linnaean Society and a staunch defender of Linnaean systematics to the end, could be expected to preserve Linnaean classification wherever possible. In this entry, published in 1804, Smith does not. Linnaeus had named and classified this arum among Gynandria Polyandria in the 1753 *Species Plantarum.*[20] By the early 1820s, when his most respected colleague, the Scottish botanist Robert Brown, had already published several essays in which he challenged Linnaeus's taxonomic decisions and argued instead for a classification in line with de Candolle's Natural Orders, Smith had to concede.[21] His less drastic (albeit still thoroughly Linnaean in its evaluation) revision two decades before of another Linnaean classification and name suggests how the taxonomic ground had begun to shift even in England by 1800.

The publication history of later editions of the Sowerby/Smith *English Botany* emphatically illustrates the effect of the shift from Linnaeus to the "natural orders" on a work that was historically identified with Smith and Linnaean systematics. First published in thirty-six volumes between 1790 and 1814 with thousands of engraved illustrations by James Sowerby, and supplementary "remarks" by Smith, *English Botany* was enormous and it was expensive. It was heavily illustrated and each illustration was hand colored. Successive volumes of its first edition appeared more or less every year. In 1814, Sowerby published a series of *Indexes* for the entire set, which allowed readers to find plants by various means: in order of their Linnaean classification or alphabetically by Latin as well as English common names.[22] In 1863, the work was republished by Robert Hardwicke. In this "enlarged" third edition all entries were rearranged to match the new systematics based on the natural orders. At about the time this reorganized edition appeared in print, the owner (tentatively identified as Charles Jenner) of one copy of the 1790–1814 edition pulled the thirty-six volumes apart to organize the plates and accompanying commentary according to their natural orders. To Sowerby's 1814 *Indexes,* this owner added in manuscript a general table of contents for the newly re-ordered volumes. For each volume, he supplied a specific list of contents and renumbered the thousands of plates on front sheets by hand, though not the plates themselves, which Sowerby had numbered at the time he engraved them.[23]

Late-eighteenth-century and Romantic attention to taxonomy as a vehicle for asking questions about parts and wholes, the problem of species, and alien or monstrous kinds prompted both specific inquiries and a ready set of analogues for thinking about human behavior. In Edmund Burke's *Reflections on the Revolution in France*, taxonomic distinctions satirize categories of dissent. Commenting on the Reverend Richard Price's admiration for the "great company of great preachers," Burke sneers: "It would certainly be a valuable addition of nondescripts to the ample collection of known classes, genera and species, which at present beautify the *hortus siccus* of dissent."[24] One impetus for invoking taxonomic categories in this fashion—there were surely many—was Erasmus Darwin's *Botanic Garden*, in which personified plants illustrate, to the dismay of some readers, the sexual proclivities of the Linnaean system.[25] Darwin's easy brief for slipping from plants to humans reappears in Mary Wollstonecraft's *Vindications of the Rights of Woman* (1792), perhaps prompted by the drawing lessons in botany she took with Sowerby.[26] In the section of this essay that refutes the claim that there is sex in souls and thus that women have a weaker sex and virtue, Wollstonecraft stages for woman a remarkable theft of the "masculine" part in the Linnaean sexual system: "The stamen of immortality, if I may be allowed the phrase, is the perfectibility of human reason; for, were man created perfect, or did a flood of knowledge break in upon him, when he arrived at maturity, that precluded error, I should doubt whether his existence would be continued after the dissolution of the body."[27] Inasmuch as the ironic counterpoint to these "if" and conditional clauses is Wollstonecraft's sustained argument from cases throughout this *Vindication* that men are not created perfect, her use of "stamen" to make an overarching claim about immortality and reason for both sexes is daring indeed. Later in this *Vindication*, she uses a taxonomic analogy between women trained to be dependent and spaniels to suggest, under the banner of Burkean "custom," an evolutionary downward spiral for women and the spaniel's ears:

> Servitude not only debases the individual, but its effects seem to be transmitted to posterity. Considering the length of time that women have been dependent, is it surprising that some of them hug their chains, and fawn like the spaniel? "These dogs," observes a naturalist, "at first kept their ears erect; but custom has superseded nature, and a token of fear is become a beauty."[28]

I offer these snapshots of how botanical and taxonomic discussions could and did prompt such analogy to suggest that this slippage is itself analogous to the way Charlotte Smith and John Clare invent poetic figures to map the implications of species and taxonomic names. If this poetic activity evidently owes a good deal to the material world of plants and, to a lesser extent, that of birds, its work is patently not material but poetic and, if my adjacent evidence suffices, oddly cultural as well. These argumentative leaps from plants to botanical taxonomy to cultural analogues and finally to evidence of poetic agency are, in the end, very broad indeed. Between each of them is at least a gap, perhaps an abyss, where live large differences in kind. So be it. Without these leaps, the poetic figuration of Smith and Clare stands, as it were, on the margins of Romantic poetics, like local curiosities that lack a certain purchase on the field of study we call Romanticism. My argument here is that what these poets do with such figures has, after all, much to do with the abiding questions of Romanticism.

If Wollstonecraft made her own lessons out of what she learned from Sowerby, and Leigh Hunt paid close poetic attention to his sister-in-law Elizabeth Kent's published books on botany,[29] Smith and Clare probably knew much more about natural history. Both wrote poems that deal extensively with natural history, with special notice of botany as well as ornithology. Both comment tacitly or directly on the taxonomic distinctions and debates of the era, although their strategies for handling taxonomic information could hardly be more different. Whereas Smith uses Linnaean nomenclature and English botanical authorities in part to show that she knows whereof she writes, Clare scorns Linnaean names in favor of a common nomenclature and local taxonomy that seek to reinstate John Ray's nomenclature. What these poets do with taxonomy is finally and oddly resistant to Romantic botanical culture. Each invents figures that trope on taxonomic distinctions in the sense that they turn away from a literal, botanical meaning. More precisely, they turn from a meaning that is putatively literal by making its figurality manifest. If such figures are part of a "material" culture, it is only by a poetic *via negativa* that disturbs what Pierre Bourdieu characterizes as the habitus[30]—that set of values or predispositions that create an ideology or culture—by twisting a cultural marker so that it can do the work of poetic agency.

For each, the resistance is different. Smith's resistance is diffuse, at times muffled; at other times it appears to target a system of binomial values modeled on and triggered by Linnaean nomenclature. Perhaps because the Linnaean system was still relatively secure in England at the time of Smith's death in 1806, she gingerly takes aim at the taxonomic authority she meticulously acknowledges in notes. Writing in the 1820s, Clare takes quite explicit aim at Linnaean nomenclature, on the grounds that it had replaced common names with Latin ones and made reproductive organs the basis for classification with little taxonomic regard for other morphological aspects, which Clare identified in the local varieties.

Smith's knowledge of natural history is well displayed in poems and pedagogical essays, the latter typically in the form of a dialogue, in *Minor Morals* (1798) and *Sketches of Natural History* (1798), *Conversations introducing Poetry* (1804), and the posthumous *A History of Birds* (1807), the last of her books for children. In one poem she lists several species of *Erica* (i.e., heaths) (*Poems* 198, l. 18)—a genus whose traits and species are frequently discussed and illustrated in the botanical literature of the period.[31] In a note to another she reviews a long-standing debate about the nomenclature and classification of plants as Geranium or Pelargonium.[32] Smith's poetic nomenclature is tactfully split between common names—which usually appear in the text—and Linnaean/latinate names, which she appends in notes. The only exceptions to this practice appear to be a few genera that were, by 1800, so frequently illustrated and identified with their Latin names that Smith chooses instead to put that name in the text and the common name in a note, such as clematis and arum, the latter a Linnaean genus whose classification and species were discussed and frequently illustrated in botanical magazines and books. In the event that the reader misses these cues, Smith's notes cite numerous works on botany.³³

So much Englishness and seeming deference to Linnaean nomenclature may be strategic. Like other women writing about botany during the 1790s and shortly after, Smith may have chosen to present her knowledge of botany with as much decorousness and conventional moral fervor as seemed to be requisite after the Reverend Richard Polwhele's 1798 broadside attack in *The Unsex'd Females* on women writers in general and, in particular, on those who taught the prurient science of botany to boys and girls.³⁴ Erasmus Darwin's *Botanic Garden* had already supplied the lubricious foundation for Polwhele's indictment of the "bliss botanic" among the sexual horrors women writers on botany for children could be expected to encourage.³⁵ As a woman writer and a teacher of natural history to children, Smith fit both profiles in her published books, including one that appeared in the same year as Polwhele's diatribe. Alan Bewell has observed that Polwhele echoes and focuses a decade of mounting criticism in the press throughout the 1790s against a constellation that included women, dissenters, and radical botanists like Darwin and signaled, in effect, the hold of radical, even Jacobin, principles to an anti-Jacobin government and press.³⁶ Given the historical moment, Smith's several citations of Erasmus Darwin merit respect (*Poems* 287 n.).

In "A Walk in the Shrubbery" (a title that looks like the occasion for a Monty Python send-up of the English penchant for natural history), Smith presents herself demurely as a "moralising botanist" who uses no personifications or pointed notes, strategies that Darwin had made so familiar in the view of his critics as to be salacious (*Poems* 303). To be sure, she is careful not to return to the highly sexualized scenes of his *Loves of the Plants*. But Smith's moral dicta were not always politically orthodox, not even in her books for children. Even in the *Elegiac Sonnets*, she could risk her claim to public favor by using "the rights of man," an incendiary, pro-revolutionary phrase by 1792, to comment bitterly that the death of a beggar signals the "insulted rights of man" (*Poems* 97). Across several genres, Smith's writing frequently invokes taxonomic vocabulary to assist ethical arguments. She begins to make extended use of poetic figures grounded in botany, and other branches of natural history occur in *The Emigrants* (1793), which urges its English audience to imagine French émigrés as not monstrously other but as beings whose sufferings, the speaker urges, strongly resemble the suffering of the English poor, as well as her own.

In part 1 of *The Emigrants*, Smith invokes the language of taxonomy to prompt readers' sympathy for "affliction's countless tribes" (*Poems* 137, l. 64). Since the end of the seventeenth century, "tribe" has been variously used to refer to classes of plants and animals as well as human groups. Its usage as a classificatory term is instructively labile from 1640 until the mid and late nineteenth century, when "tribe" settled down to mean a level of taxonomy inferior to order but superior to genus. Until then, it could mean a very general class, as in "the two grand tribes of vegetables are acid and alkaline" or a "tribe of birds whose habit it is to fly as a flock"; a species, or an order that contained a number of genera, roughly equivalent to Antoine de Jussieu's taxonomic use of the term "family" in 1789 to supplant the Linnaean genera as the largest category and division in plant and animal classification.[37] Smith's poetic deployment of the term elsewhere in the poem suggests that she knows precisely how to inflect this term such that its lability might serve an argument that makes taxonomy support a larger political argument.

In part 2 of the poem, Smith concludes an extended description of the flowers and birds of her native Sussex with a sharp critique of recent violence in Revolutionary France. Turning back to those who inhabit "this sea-fenc'd isle," the speaker pointedly asks: "What is the promise of the infant year to" those who watch France and to "the Man, who thinks,/Blush for his species?" (*Poems* 151, ll. 62–68). The syntax, undecided between its singular and plural referents, urges in the language of taxonomy Smith's claim that English readers should recognize their solidarity with other members of their "species." The trajectory of Smith's nomenclature from "tribes" to same "species" nudges the English reader to recognize the French not as some other tribe, but as the same species. The taxonomic lability of the antecedent "tribe" during Smith's lifetime implies that like families, tribes may, as Jussieu argued, have affinities between as well as within these groups.

The taxonomic variability Smith implies here and elsewhere in her poems persistently works a relay between the minutiae of natural history and the human moral lessons those minutiae illustrate. This strategy is especially evident in *Minor Morals*, a series of dialogues between an aunt and her nieces and nephews in which the microscopic observation of botanical subjects prompts the putatively "minor morals" of her title. In the first dialogue, the aunt explains how the underside of a moth looks under a microscope—"an infinite number of feathered quills" more exquisitely detailed than "the finest miniature."[38] She then observes, with an air of digressiveness that I read as part of her rhetorical cover, that the study of botany wards off indolence. The dialogue closes with a story about other children who, having returned from a British colony in the care of a black slave, are at once demanding and dependent little tyrants whose characters have been warped by indolence. The moral scrutiny Smith invites with this tale would translate the minute inspection of a moth into an equally fine moral microscopic capable of recognizing how—to give the implicit moral of this tale its largest geopolitical argument—slavery and servitude weaken the children of masters.

The indirection that here marks Smith's drawing of her moral also guides the narrative logic of "To the fire-fly of Jamaica, seen in a collection," a poem she first published in *Conversations introducing Poetry* (*Poems* 204–07). In mid-poem the speaker's meditation on an impaled firefly shifts, first to a slave captive who cannot escape and so cannot welcome this firefly to light his way, then to a Naturalist (so capitalized), and finally to a meditation on the difference between Ostentation and Friendship (also so capitalized). The poem is a curious poetic performance, encased in intricate rhyme schemes and stanzaic patterns that are, far more than unrhymed iambic pentameter, Smith's hallmark. Although these formal devices call attention to the surface of the argument, the analogies the poem draws between the firefly, the slave, and the Naturalist work, as it were, under poetic cover, but with an assist from Smith's notes, in which Latinate names for flora and fauna alternate with notes that describe the suffering of slaves in the colonies.

The poem's concluding stanza about the vanity of ostentatious display and loyal friendship, surely a charge to be laid at the feet of the Naturalist and not the firefly so pinned, scats away from the more politically charged moral embedded in the speaker's odd narrative relay. Whereas some anti-abolitionists claimed that Africans belonged to another, lesser species, Smith's taxonomic distinctions are offered without a difference for individuals: whether a dead firefly, a slave, or a Naturalist, all are linked in an odd and oddly sympathetic chain of being that leads, obliquely enough, to a closing moral about the vanity of ostentatious display. This lesson looks like a cover to deflect attention from the poem's less overt notice of commonalities that work against using taxonomic classification to pin others to a grid.

"To a Geranium," also included in Smith's *Conversations*, concerns a species of geranium that is native to Africa but, once "naturalized in foreign earth" (*Poems* 209, l. 21), blooms during the winter. Smith's extended note about geraniums and pelargoniums obliquely indicates the outlines of an argument that goes beyond botanical enthusiasm for naturalized exotic flowers. She observes that among the numerous varieties of geranium (or pelargonium—take your pick) are both native and exotic species. The latter are "crush'd" by "Caffres" in South Africa, but survive first in the brilliant luxury of an English conservatory, and then in the home of the speaker, to cheer her "like friends in adverse fortune true" (*Poems* 209, l. 28). "To a Geranium" is not an antislavery poem; indeed, its brief notice of South African "Caffres" who crush exotic blooms may be offered in unflattering contrast to those English botanists who cherish them.

Yet Smith's phrase "naturalized in foreign earth" runs against the grain of this implied contrast. The "foreign earth" here is English soil and hearth, which have until this line constituted the perspective from which the speaker offers her admiration for exotic geraniums in English hothouses. Although slightly marked, this shift in perspective works in tandem with the global vantage point suggested by the note, in which the genus Geranium is presented as inclusive of native and exotic species. Like *Minor Morals*, these conversations between one Mrs. Talbot and her children, at once lessons about natural history and about poetic forms and usage, imagine these enterprises as both taxonomic and figurative markers for a moral order in the forms of knowledge they offer.

Smith's "Flora," first published in *Conversations introducing Poetry* (1804) and reissued in the posthumous *Beachy Head, Fables, and Other Poems* (1807), presents an array of personified insects and flowers who guard Flora, among them tiny sylphs, called "pigmy warriors," who use lichen fronds as shields and weapons (*Poems* 281–82, ll. 50–78). Among the trees that recognize Flora's "soft sway" over them is the Oak, whose "giant produce may command the World" (*Poems* 284, ll. 132–34). Stuart Curran's note specifies the allusion here: oak was the main-stay of English naval shipbuilding, a growth industry throughout the 1790s and the Napoleonic Wars (*Poems* 284 n.).

I wonder about this military strain in a poem whose luxurious botanical cat-alog is seducing, perhaps artfully distracting. Like *The Emigrants*, which is pointedly dated November 1792, two months after the September Massacres, "Flora" was published at a time when English naval vigilance against the French navy was a constant topic in the press and in private, with particularly anxious de-bate about possible French landings on English coasts, among them the Sussex coast. Placed beside hints in *The Emigrants* and *Beachy Head*, poems that specify natural and geological histories as exempla that urge some future reconciliation be-tween England and the Continent to which it was once joined, the miniature, then giant, battle array of Smith's "Flora" may pick up the thread of an argument that runs aslant the poem's more apparent brief for, put simply, the realm of Flora.

The poetic work of this realm has everything to do with contemporary debates about botanical taxonomy and nomenclature. Indeed, in this regard Smith seems to have given Clare's poetic practice a nudge. In the first of his several letters on natural history, he interweaves a rambling taxonomic discussion of common plant names with sharp-eyed praise for Smith's poem "Flora," from which he quotes these lines: "Amid its *waving swords*, in flaming gold/The iris towers" (*Poems* 285, ll. 156–57; Clare's italics in Clare, *NHPW* 20). Other plant and bird names that surround these lines in Smith's poem suggest the logic of Clare's poetic interest in those "waving swords." The same and immediately preceding stanzas list "Adder's tongue," also called "Hart's tongue," "Reed bird" (in fact a bird), "Arrowhead," and "Crowfoot." In a note, Smith explains the Latin name *Hypericum* by quoting William Cowper's description in *The Task*: "Hypericum all bloom, so thick a swarm/Of flowers like flies clothing her slender rods/That scarce a leaf appears."[39]

Smith's roster of hybrid common names—hybrid in the sense that they cross from one phylogenetic kingdom to another—is hardly unique to "Flora." "The Horloge of the Fields," a botanical survey of flowers that close and open at different times of day, mentions "Goatsbeard" (*Poems* 297, l. 38). "Wheat-ear," the title of a poem in *Con-versations introducing Poetry*, is a bird's name with a double cross to plants and human anatomy (*Poems* 194–96). The most arresting example is retrospectively suggested by

Clare's letter, which turns to Smith in the midst of a brief discussion of *Orchis* species. In the poem Smith published first in *Minor Morals* under the title "The Kalendar of Flora," then republished in *Conversations introducing Poetry* as "Wildflowers," the botanical catalog includes "the Orchis race with varied beauty charm, [to] mock the exploring bee, or fly's aerial form" (*Poems* 191, ll. 23–24). Treated as a plural subject, "Orchis race" includes both the species she names in a note, the "Fly and Bee Orchis" (Figures 8.3 and 8.4), and other species, including one whose Linnaean classification even Sir James Edward Smith had disputed in *English Botany*. In the same poem, she lists another plant, *arum maculatum*, whose Linnaean classification Sir James Edward Smith also challenged.[40] In *Beachy Head*, Smith is more expansive about the taxonomic and poetic interest of the *Orchis* genus, noting its

> flowers,
> Of wondrous mockery; some resembling bees,
> In velvet vest, intent on their sweet toil,
> While others mimic flies, that lightly sport
> In the green shade. (*Poems* 235–36, ll. 443–47)

Her note to these lines echoes Sir James Edward Smith's commentary in *English Botany*: "Linnaeus, misled by the variations to which some of the tribe are really subject, has perhaps too rashly esteemed all those which resemble insects, as forming only one which he terms Ophrys insectifera."[41] Insofar as many common names in the nomenclature of natural history perform similar double or triple crosses, they are a material and accidental gift to poets. In Smith's poetry, the taxonomic wit of such figures disables or helps to disable rigid binomialism—French versus English, beggars versus the reader.

Nearly two decades after Smith's death, Clare admired her botanical figures as instances of his own preference for common plant names, such as "Ragged Robbin," over Linnaean ones (Lychnis in this instance).[42] Finding in the seventeenth-century Ray authority for this preference, Clare defended common nomenclature as part of a larger brief for dialect words and local knowledge. Both constituted for Clare essential signs in language of a history and culture that were likely otherwise to slip away, pushed aside by a homogenizing language drift.[43] From this perspective, Linnaeus's Latin nomenclature was devastating because it categorically refused local names and instead imposed its own, distinctly un-English, set of mutually self-contained names. By contrast, common names carry their human history, as it were, on their backs. In a note to his *Shepherd's Calendar*, for example, Clare explains that the word "fumitory," a mixture derived from "red and purple mottled flowers" then boiled in whey and milk, is used by farm women to bleach their complexions and that "cheese" is a local term for the crushed seed of the mallow.[44]

Clare's letters and manuscript plant lists document his extensive knowledge of the local natural history of Helpston, to which he added in the 1820s by

Figure 8.3. James Sowerby [and James E. Smith], *English Botany, 36 vols.* (London: J. Davis, 1790–1814), Vol. 6, Pl. 383: Bee Orchis (*Orphrys apifera*). Hand colored engraving by Sowerby from his own drawing. Courtesy of the Linda Hall Library, Kansas City, MO.

64.

Figure 8.4. Sowerby, *English Botany* Vol. 1, Pl. 64: Fly Orchis (*Orphrys muscifera*). Hand colored engraving. Courtesy of the Linda Hall Library, Kansas City, MO.

exchanging specimens, information, books, and articles with two artisan naturalists, Edmund Tyrell Artis, an archaeologist who published work on fossil plants,
and Joseph Henderson, a gardener who published a paper on ferns. Both were servants at Milton Hall, whose proprietor Lord Fitzwilliam (the title he assumed later
in life) was also an avid naturalist.[45] Although artisan networks like this one were
certainly not classless, they gave working-class botanists such as Job Legh in Elizabeth Gaskell's *Mary Barton* a circle in which to gain some authority over what Noah
Heringman has identified as "forms of [Romantic] knowledge" about natural history.[46] To the extent that Clare was interested in local varieties, his knowledge of
natural history resembles the folk taxonomies Atran describes as being concerned
to identify all the flora of a region, but not to assimilate that knowledge with a
wider or global variety. And yet, Clare's plant lists, which included taxonomic sub-
headings, imply his persistent attention to taxonomic debates about those plants
he knew well. He was especially interested in orchis varieties, in part because he
watched them gradually disappear as Helpston land was enclosed. Because Artis
had extensive familiarity with the orchis genus or family, Clare apparently asked
him to write the species names in a list that is included among Clare's own lists in
the modern edition of his writing on natural history.[47]

Because Clare self-consciously hoarded local dialect words whose variety cor-
responded in a formal sense to the variety he also found in the natural world, he
was not above taking fairly crass aim at the Linnaean system. In the poem, "A
Ploughmans Skill at Classification after the Lineian Arrangement," he makes that
arrangement the template for marital bickering between a ploughman and his
haranguing wife. Exasperated with her complaints, he concludes that if he is a
hog, she is a sow.[48] So caricatured, the sexual basis for Linnaean classification
leads to a linguistic as well as marital dead end in which both husband and wife
produce bitter, self-canceling taunts about men and women. Much as Erasmus
Darwin's notice of hermaphroditic plants emphasizes the sexual economy of Linnaean systematics, so do Clare's remarks about hermaphroditic plants in his
letters on natural history query Linnaeus's conclusions about which plants are
hermaphroditic and which are not.

In an early poem, "Recolections after a Ramble," Clare specifies the poetic
resources made available by common botanical names:

> Some went searching by the wood
> Peeping neath the weaving thorn
> Where the pouchd lip'd cuckoo bud
> From its snug retreat was torn
> Where the ragged robbin grew
> With its pipd stem streakd wi jet
> And the crow flowers golden hue
> Carless plenty easier met.[49]

There is nothing careless about the poetic strategy at work in these lines as names of birds (or what look at first like names of birds) morph into names of flowers. As a survey of his bird poems makes clear, Clare was enormously knowledgeable about ornithology. The fact that his spelling is notoriously nonstandard has led at least one naive reader astray. Once that reader figured out that she had made a category error—been tricked as it were into a catachresis inasmuch as this cuckoo hidden beneath a thorn tree is no thorn bird but a flower—she was forced to slow down and read for detail. The next two "bird–flower" descriptions are poised to attract a reader who is by now in the know—that is to say, in the grip of precisely the kind of particular knowledge about the real world, Clare's world, that this poet urges.[50] In a natural history letter on birds, he praises Smith for using similar bird names in a sonnet about a bird named "the fern Owl or Goat Sucker or Night jar or night hawk." Noting that the fern owl and nighthawk differ in key details, he also lists the names for a related species: "Hay chats straw chats nettle chats &c" (*NHPW* 108).

Asserting that common names crowd the imagination with poetic associations as Latin ones tied to Linnaean classification do not (*NHPW* 41), Clare defends two species whose common names "cross" birds with plants, the "pouchd lip'd cuckoo bud" of his poem "Recolections" and the "blue flowered cranes bill," against their Latinate classifications (*NHPW* 15, 22). The "cuckoo" flower is, as he notes, a variety of orchis that is "found in Spring with the blue bells." Its flowers are, he goes on to explain, "purple & freckld with paler spots inside and its leaves are spotted with jet like the arum" (*NHPW* 15–16). The most likely candidate for this orchis is *Orchis mascula* (Figure 8.5), identified in the Sowerby/Smith *English Botany* as "Early Purple orchis," although this species is frequently found hybridized with *Orchis morio* (Figure 8.6), which Sowerby and Smith call "Green-winged meadow orchis."[51]

Clare's poetic choice, like Charlotte Smith's nearly a generation earlier, is hardly accidental. The Orchis "family" or genus belongs to the natural order of the *Orchidacaeae*, a group that puzzled botanists as they tried to match local and exotic species with the Linnaean classification. Clare mentions orchis varieties in this and another letter on natural history, in his journal, and in two manuscript plant lists of orchid names and locations (one of them in Artis's hand) (*NHPW* 295–302). Orchid illustrations—from the presentation of orchid varieties in Curtis's *Botanical Magazine* and Sowerby/Smith's *English Botany* to Francis Bauer's detailed, exquisite drawings of orchid anatomy and James Bateman's at-least-as-large-as-life presentation of the exotic orchids in his *Orchidaecae of Mexico and Guatemala*—amply document the effort to resolve classificatory relationships within this group.[52] Terrestrial orchids like Clare's "cuckoo flower" belong to the genus Orchis. Because this genus includes the largest number and the most common European species, its name was eventually given to the entire order.[53] Clare's "large blue flowerd cranes bill or wild geranium" also belongs to a genus whose name botanists debated into the early decades of the nineteenth century, a taxonomic dispute also available to Clare via Smith's poems. Clare will only go so far

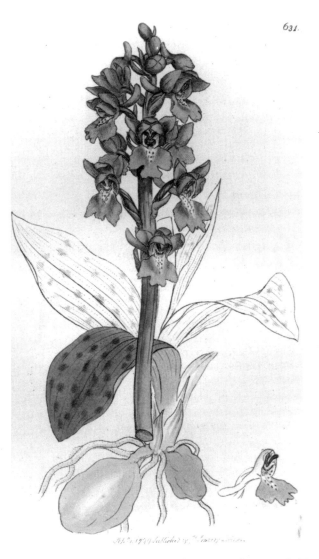

631.

Figure 8.5. Sowerby, *English Botany* Vol. 9, Pl. 631: *Orchis mascula* (Early Purple Orchis). Hand colored engraving. Courtesy of the Linda Hall Library, Kansas City, MO.

2059.

Pub. in aug. published by J.d Sowerby London

Figure 8.6. Sowerby, *English Botany* Vol. 20, Pl. 2059: *Orchis morio* (Green-winged Meadow Orchis). Hand colored engraving. Courtesy of the Linda Hall Library, Kansas City, MO.

Figure 8.7. Curtis, *Flora londinensis* Vol.2, Pl. 49: *Geranium pratense* (Meadow or Blue Flowered Cranesbill). Hand colored engraving. Courtesy of the Linda Hall Library, Kansas City, MO.

as to call it a member of the "geranium tribe." His modern editor identifies this plant as geranium pratense or "meadow cranes bill" (*NHPW* 22) (Figure 8.7).

Some of Clare's common names and taxonomic distinctions reappear in the botanical systematics based on Natural Orders to which English botanists finally transferred their allegiance in the late 1820s. *Ladies' Botany*, one of John Lindley's popular, and frequently reprinted, books on botany, lists among the species of the Orchis tribe common wild plants with names such as Bee Orchis (the base of its white flowers look like a bumblebee or yellow jacket), Fly Orchis, "whose lips," Lindley declares, "resemble the insects after which they are named," as well as "Man Orchis, Lizard Orchis, Butterfly Orchis, Bird's Nest Orchis."[54] Lindley wrote this and other works for popular audiences, but he was an eminent botanist, known for his work on orchids and for his early popular explanation of the Natural Orders.[55] The figural punt of Lindley's botanical description of the Fly orchis scores one for poetic figuration. In the much later *Treasury of Botany*, which Lindley began but others completed after his death, more names were added to the list: "Crane Fly Orchis," "Dog Orchis," "Medusa's Head Orchis," "Monkey Orchis," "Frog Orchis," "Green-Man Orchis," among others.[56] Most of these names had also appeared in the 1790–1814 *English Botany*, which Charlotte Smith used to indicate the poetic and taxonomic interest of common plant names whose Linnaean classification earlier English botanists had disputed.[57]

Clare's "pouchd lip'd cuckoo bud" is absent from these lists, though its linguistic affiliation to all these common names for species is apparent. Lindley also announces that the geranium–pelargonium debate is over: the name now assigned to the genus is the more familiar one Clare had grudgingly used—geranium. Clare's sense of why Linnaean names would not do is, as the catachresis of his "Recolections" makes clear, grounded in poetics. In an 1823 letter to his publisher James Hessey, Clare included a verbal sketch of an essay on poetry that he hoped to publish in the *London Magazine* but never did. A highly polemical attack on the artificial diction and aristocratic posturing of Renaissance pastoral poetry, the sketch includes an especially sharp riff on the flower names the artificed nature of this poetry is said to prefer: "instead of the wild flowers of the wilderness growing [on] its brink it [nature] was dizened out with daffodils and narcissus lilys &c &c." Clare delivers the botanical and class subtext for this complaint in a note: "it was too vulgar to call things by their own names—so that the dead nettle had the good fortune to get nighted with the fine title of 'archangel' by some poet of this 'golden age' & the nettle ought to be thankful for his lordships pastoral condesencion—such things may be thought wearisome trifles to mention but it is mentioned to illustrate a trifling subject."[58]

Clare's pun on "trifles" is dead serious. Far from being a mere trifle, the choice of flower names in Renaissance pastoral trifles with the "subject" at hand. Precisely because it moves without marking the difference from artificial, not real, pastoral poems to flowers that are not from a native English pastoral landscape, Clare's note insists that the "subject" being trifled with is at once poetry and botany, much as his attraction to Charlotte Smith's botanical figures registers his understanding of how common names and poetic figures might be made to occupy the same ground.

The quirky alliance between taxonomic inquiry and Romantic poetics that I have sketched here suggests one way to think about placing Romantic poetic subjectivity in relation with Romanticism's quasi-"material" cultures. For if poetic figures can be said to specify the stubborn as well as playful resistance that constitutes the work of embedding self in world, this project is as much Romantic as the claim to be outside time or community in the sense offered by Coleridge's philosophical spectator "*ab extra*"[59] or the mariner of his *Rime of the Ancient Mariner.*

It is also of a piece with botanical efforts to imagine the synecdochic relation between this whole and its parts—an effort made all the more necessary and difficult as more species became known to English and Continental botanists. Both inquiries ask questions that haunt modernity and, within that extended domain, Romanticism: what is at stake in all acts of representation, whether political, literary, or scientific? In what sense can individuals or species, texts, and particulars be said to belong to a larger body, be it nation, culture, or philosophical idea? What is exemplary and what is not? What belongs to us, what does not? By considering how these questions are addressed in the name of natural history, particularly botany, systematics, and nomenclature, we are better positioned to notice how this very material part of Romanticism is itself caught up in the work of thinking about representation. By making the underlying figurality of taxonomic distinctions and nomenclature manifest, these questions also show how individual agents work on the habitus of Romantic botanical culture by resisting as well as incorporating taxonomic distinctions and schemas.

This argument invites application to another set of questions about current disciplines of critical understanding concerning Romanticism. Put briefly, those questions concern the point of intersection between theories of the subject and a disciplinary approach whose ground tends to be inflected by cultural critique—the investigation of Romantic culture as a site where Romantic individuality and subjectivity exist mostly on sufferance, regarded by many as suspect terms that stand in the way of cultural work that is more pressing. This essay argues against such bifurcation by casting a line (bridging a gap) between Romantic botanical culture and the work of two poets. My rationale for making this problematic critical move owes some of its impetus to recent work in ethnographic theory that reexamines its earlier assumptions about collectivity in indigenous culture.[60]

As a key historical moment in modernity's effort to specify what is at stake in all acts of representation—be they political or literary or both of these and more—Romanticism offers ample evidence of the extent to which its writers and other actors tried to understand terms such as *individual, collective, representative* or *exemplary, singularity* and *difference, periphery* and *center, self,* and *nation* and *other*. Antoine de Jussieu's recognition of interfamilial as well as intra-familial affinities in classificatory systems suggests one figure for the disciplinary aim I have in view.[61] If Romanticism was a culture whose attention to subjectivity marks its poetic practices, it was also one in which competing strands of social and collective understanding vied for prominence. My sense of how critical practice might acknowledge this double bind is indebted to recent work by the French anthropologist Marc Augé, who argues that witchcraft and divinatory practices among the Lagoon peoples of the Ivory Coast give outsiders opportunities to situate themselves in communities from which they might otherwise be excluded as distinctly, stubbornly "other." Such practices, Augé suggests, recognize in persons "the existence of an individual sign, a singular formula, a residue not exhausted by all the combinations of filiation and alliance."[62]

What if we were to understand Romantic writing as capable both of marking its relation to its culture and time and specifying a persistent, stubborn "residue" of difference, of individual agency, within that culture? This essay offers Romantic botany and the poetry of Charlotte Smith and John Clare as twin signs of that residue. From this disciplinary perspective, the study of Romanticism's cultures might well seek out flickers of individual difference as textual residues or remainders that can be read as flickering signs of the productively vexed relation between individuals and cultures. From this critical vantage point, contingent relations between literary texts, "material" cultures, and theorizing about wholes and parts have everything to do with Romanticism and modern criticism. For by proceeding in this way, literary and cultural studies are better poised to recognize how Romantic writing sediments itself in its culture but also specifies its differential, because nonmaterial, presence there. My interest in the culture of Romantic botany is driven, then, less by its taxonomic details or the brilliance of botanical illustration (although these too have their beauties) than by their status as substrata out of which Romantic poets craft poetic singularities.

Notes

The epigraph that opens this chapter is from *Essays on Philosophical Subjects* (Dublin: Wogun, Byrne, Moor et al., 1795), 16-17. Eric Gidal quotes part of Smith's remark in *Poetic Exhibitions* (Lewisburg: Bucknell UP, 2001), 54, in his instructive discussion of the eighteenth-century concern to organize the array of curiosities first exhibited at the new British Museum.

1. Quoted by Edward Edwards, *Lives of the founders of the British Museum: with notices of its chief augmentors and other benefactors, 1570–1870* (London: Trübner, 1870),

296–97; and by Gidal, *Poetic Exhibitions*, 43. For an overview of Sloane's role in the establishment of the British Museum, see Arthur MacGregor, ed., *Sir Hans Sloane: Collector, Scientist, Antiquary, Founding Father of the British Museum* (London: British Museum, 1994).

2. Francis Hutcheson, *Inquiry into the Original of Our Ideas of Beauty and Virtue*, 4th ed. (London: Midwinter, 1738), 31, quoted in Gidal, *Poetic Exhibitions*, 53. Although the naturalist Thomas Pennant expressed reservations about Linnaeus's binominal nomenclature, he admired Linnaean systematics for its management of the "unmanageable profusion of names" that had characterized earlier systematics. Pennant, *History of Quadrupeds*, 2 vols. (London: B. and J. White, 1793), 1:157; quoted by Harriet Ritvo, *The Platypus and the Mermaid and Other Figments of the Classifying Imagination* (Cambridge, MA: Harvard UP, 1997), 51. Thomas Martyn offers a dissenting opinion late in the eighteenth century, arguing for the "sublime disorder" of nature and the necessity to "emancipate ourselves from system." Martyn, *A New Dictionary of Natural History*, 2 vols. (London: Harrison, 1785), 1:no pag. The anonymous *Dictionary of Natural History* (London: C. Whittingham, 1802), iii, echoed Martyn's sentiments. Quoted by Ritvo, *Platypus and the Mermaid*, 24.

3. Siegfried Kracauer, *History: The Last Things Before the Last* (New York: Oxford UP, 1969), 107–27.

4. David E. Allen, *The Naturalist in Britain* (London: Penguin, 1976), 6.

5. Smith, *Essays*, 3.

6. Bill Brown, *The Material Unconscious* (Cambridge, MA: Harvard UP, 1996), 5.

7. David L. Clark, "Kant's Aliens," *The New Centennial Review: borders/america* 1 (2001): 201–69.

8. A significant number of botanists, artists, and naturalists were Quakers, among them William and John Bartram, John Fothergill, James Backman, James Lee, and Sydney Parkinson, the artist for James Cook's *Endeavour* whose illustrations are the basis for Joseph Banks's massively illustrated *Florilegia*. Other botanists, including James E. Smith, were Unitarian; like many botanists of this period, Robert Brown was a Scotsman. He also came from a strong and dissenting Presbyterian family. Ann Shteir reviews the careers of several women botanists in *Cultivating Women, Cultivating Science: Flora's Daughters and Botany in England, 1760–1860* (Baltimore: Johns Hopkins UP, 1996). For an overview of these and other botanical careers, see relevant entries in Ray Desmond's *Dictionary of British and Irish Botanists and Horticulturalists* (London: Taylor and Francis, 1994); and Wilfrid Blunt and William T. Stearn, *The Art of Botanical Illustration*, rev. ed. (Kew: Antique Collector's Club, 1994).

9. At Banks's request, William Bligh took breadfruit saplings from Tahiti on board the *Bounty* to carry them to the West Indies, where they were to supply a needed food source to African slaves on British plantations, who were starving and dying at the rate of fifteen thousand a year. Years later, after the famous mutiny and Bligh's eventual return to England and legal vindication, Banks again asked him to bring breadfruit to the West Indies. This time Bligh successfully completed the commission, but West Indian/African slaves did not like them. See David Mackay, "Agents of Empire: The Banksian Collectors and Evaluation of New Lands," *Visions of Empire*, ed. David Philip Miller and Peter Hanns Reill (Cam-

bridge: Cambridge UP, 1996), 47; and John Gascoigne, "The Ordering of Nature and the Ordering of Empire: A Commentary," *Visions*, 107–116.

10. For an overview of these movements of plants, see Mackay, "Agents of Empire." For a fictional return to the nineteenth-century transportation of fruit trees to Australian settlers, see David Malouf, *Remembering Babylon* (New York: Pantheon, 1993).

11. Robert Thornton, *Temple of Flora* (London, 1799–1807), Pl. 22; James Sowerby [and Sir James Edward Smith], *English Botany*, 36 vols. (London: J. Davis, 1790–1814), 19:1298. The artist James Sowerby was and is regarded as the author of *English Botany*. Sir James Edward Smith provided the commentary. Cf. John Clare, *Natural History Prose Writings*, ed. Margaret Grainger (New York: Clarendon, 1983), 14, hereafter cited as *NHPW*; and Charlotte Smith, *Poems of Charlotte Smith*, ed. Stuart Curran (New York: Oxford UP, 1993), 190; hereafter cited as *Poems*. Works consulted for my Note About the Cover in this volume (identical to Figure 8.1) include Blunt and Stearn, *The Art of Botanical Illustration*, 236–43; Handasyde Buchanan, "Bibliographical Notes," *The Temple of Flora*, ed. Geoffrey Grigson (London: William Collins, 1951), 13–20; and Charlotte Klonk, *Science and the Perception of Nature* (New Haven: Yale UP, 1996), 38–65.

12. See, for example, Scott Atran, *Cognitive Foundations of Natural History: Towards an Anthropology of Science* (Cambridge: Cambridge UP, 1990); hereafter cited in the text as Atran; John Dean, "Controversy over Classification: A Case Study from the History of Botany," *Natural Order: Historical Studies of Scientific Culture*, ed. Barry Barnes and Steven Shapin (Beverly Hills, CA: Sage, 1979), 211–31; Gordon R. McQuat, "Species, Rules and Meaning: The Politics of Language and the Ends of Definitions in Nineteenth-Century Natural History," *Studies in History and Philosophy of Science* 27 (1996): 473–519; and David Hull, "Biological Species: An Inductivist's Nightmare," *How Classification Works: Nelson Goodman among the Social Sciences*, ed. Mary Douglas and David Hull (Edinburgh: Edinburgh UP, 1992), 42–68.

13. Atran, *Cognitive Foundations*, 56. Hull objects that Atran's claim for a universal disposition to identify and classify natural kinds puts too much value on folk taxonomies, which tend to emphasize a fixed or static number of species by virtue of the fact that such taxonomies are limited to one locality at a given time. From the perspective of modern evolutionary biology, Hull contends that this emphasis is irrelevant to the investigation of when and how species change in environments. From an evolutionary perspective, it is more fruitful to examine species change at the isolated periphery of a given environment—precisely the sort of information not emphasized by folk taxonomy—together with minute biological differences between sexual and asexual organisms. See Hull, "Biological Species," 55–58, 64–65.

14. Atran, *Cognitive Foundations*, 79.

15. Atran uses the term *true genera* to characterize the Aristotelian dimension of Ray's taxonomic arguments (*Cognitive Foundations* 165). For an analysis of Ray's writing on language and botany, see Jo Gladstone, "'New World of English Words': John Ray, FRS, the Dialect Protagonist, in the Context of His Times (1658–1691)," *Language, Self, and Society: A Social History of Language*, ed. Peter Brooks and Roy Porter (Cambridge: Cambridge UP, 1991), 115–53.

16. For a list of Linnaean classes and a bibliographic review of successive editions of Linnaeus's *Genera Plantarum*, see William T. Stearn, ed., *Three Prefaces on Linnaeus and*

Robert Brown (Weinheim: J. Cramer, 1962), xvi–xvii. Blanche Henrey summarizes Linnaeus's influence on botany in *British Botanical and Horticultural Literature before 1800*, 3 vols. (London: Oxford UP, 1975), 2:650–55.

17. See Lee's obituary in "James Lee," *The Gentleman's Magazine* 65 (December 1795): 1052; and James Lee, *An Introduction to Botany*, 3rd ed. (London: J. F. and C. Rivington, 1776), 19–21, 150–53.

18. Allen G. Morton, *History of Botanical Science* (New York: Academic, 1981), 287–361, 375–77.

19. Sowerby, *English Botany*, 19:1298. The more typical and modern English name, "Lords and Ladies," does not appear in Clare or the Sowerby/Smith entry.

20. Carolus von Linnaeus, *Species Plantarum*, ed. William T. Stearn (London: Ray Society, 1753), 21:964–67.

21. For a narrative of Smith's reluctant capitulation, see Margot Walker, *Sir James Edward Smith* (London: Linnean Society, 1988), 7–10, 18–21. Among Brown's significant publications are "Some Observations on the Natural Family of Plants called Compositae," *Transactions of the Linnean Society* 12 (1818): 76–150 (read to the Linnaean Society in 1816) and *Observations on the Structure and Affinities of the more remarkable plants* (London: Thomas Davison, 1826), especially 5–6, 36–37.

22. Sowerby, *General Index* (1814), *English Botany*.

23. This altered copy of the first edition of *English Botany* is owned by the Harry Ransom Humanities Research Center, University of Texas at Austin.

24. Edmund Burke, *Reflections on the Revolution in France*, ed. L. G. Marshall (Oxford: Oxford UP, 1993), 13.

25. Erasmus Darwin, *The Botanic Garden* (London: J. Johnson, 1791).

26. See Blunt and Stearn, *The Art of Botanical Illustration*, 220.

27. Mary Wollstonecraft, *Vindication of the Rights of Woman*, in *Vindications*, ed. D. L. MacDonald and Kathleen Scherf (Peterborough, ON: Broadview, 1997), 166.

28. Ibid., 204. Her source is William Smellie, *The Philosophy of Natural History*, 2 vols. (Edinburgh, 1790), 1:462.

29. Shteir, *Cultivating Women,* 142–43.

30. Pierre Bourdieu, *Outline of a Theory of Practice* (Cambridge: Cambridge UP, 1977), 78.

31. See, for example, Sowerby's *General Index* for an extensive list of *Erica* species illustrated in *English Botany*.

32. Smith, *Poems* 208 n. See as well entries in the 1817 cumulative index for early volumes of William Curtis's *Botanical Magazine*.

33. Among the botanical works listed are William Withering's *A Botanical Arrangement of English Plants*, William Mason's *English Garden*, William Woodville's *Medical Botany*, Sowerby's *English Botany*, Gilbert White's *Natural History of Selborne*, Thomas Martyn's adaptation of Philip Miller's *The Gardener's and Botanist's Dictionary*, and Erasmus Darwin's *The Economy of Vegetation* (Smith, *Poems*, 52, 54, 66, 107, 194, 210, 232, 237, 276, 295–98).

34. Richard Polwhele, *The Unsex'd Females* (London: Cadell and Davies, 1798).

35. For more discussion of how Linnaeus's anthropomorphisms invited Erasmus Darwin's "theatrical" tableaux of plant sexuality and the subsequent furor over women learning

botany, see Rachel Crawford's chapter (7) in this volume. Relationships among botanists, dissenters, and women during the Romantic period were probably deeper and broader than even Polwhele was prepared to acknowledge. Women artists and a few explorers belonged to Quaker networks of nurserymen and shipboard naturalist–artists; working-class male botanists in Lancashire met in pubs; and there existed at least one network for the amiable exchange of botanical information between dissenters and non-dissenters at Warrington Academy, where the Barbaulds, Priestleys, and others lived or congregated in the 1760s. Among the early tutors there was Jean-Paul Marat. See John H. McLachan, *Warrington Academy: Its History and Influence* (Manchester: Chetham Society, 1943), 72–84; and Anne Secord, "Science in the Pub: Artisan Botanists in Early Nineteenth-Century Lancashire," *History of Science* 32 (1994), 269–315.

36. Alan Bewell, "'Jacobin Plants': Botany as Social Theory in the 1790s," *The Wordsworth Circle* 20 (1989): 132–39.

37. See the dated uses in the *Oxford English Dictionary;* Atran, *Cognitive Foundations,* 205.

38. *Minor Morals, interspersed with sketches of natural history, historical anecdotes, and original stories* (London: Sampson Low, 1798), 20.

39. *The Task,* 6.165–67, quoted by Smith, *Poems* 283 n. See *The Poetical Works of William Cowper,* ed. H. S. Milford (Oxford: Oxford UP, 1926), 223.

40. Linnaeus, *Species Plantarum,* Pl. 1370; and James Edward Smith's commentary in Sowerby, *English Botany,* 20:1298, 1:64.

41. See Smith, *Poems* 236 n.; and Sir James Edward Smith's commentary in Sowerby, *English Botany* 1:64.

42. This plant's full modern Latin name is *Lychnis flos-cuculi.* Elizabeth Kent identifies this flower as *Galium verum* in *Flora Domestica,* a work that Clare generally praises, but with local disagreements like this one. See Elizabeth Kent, *Flora Domestica* (London: Taylor and Hessey, 1823), 232; Clare, *NHPW* 20.

43. James McKusick, "'A language that is ever green': The Ecological Vision of John Clare," *University of Toronto Quarterly* 61 (1991–1992): 235–37.

44. John Clare, *Shepherd's Calendar,* ed. Geoffrey Summerfield and Eric Robinson (Oxford: Oxford UP, 1964), 53, 56.

45. Grainger, ed., *NHPW,* xxxviii–ix.

46. By the 1840s, Gaskell could make a strong case for artisan knowledge of competing taxonomic systems. In *Mary Barton,* she notes that among the common hand-loom weavers "there are botanists . . . equally familiar with either the Linnaean or the Natural system, who know the name and habitat of every plant within a day's walk." See *Mary Barton,* ed. Macdonald Daly (New York: Penguin, 1996), 16. For a fuller account of Gaskell's novel, see Amy King's chapter (9) in this volume. A comparison of working-class or artisan networks with Banks's aristocratic, royally sponsored collection of specimens and taxonomic analysis is intriguing. Banks used Linnaean systematics to extend the logic of its inventor by centralizing and codifying information gathered from around the world into a systematic account of its flora and fauna. The Sowerby family performed a middle-class version of the Banksian enterprise. Artisan and working-class networks like that of Clare, Artis, and Henderson sometimes exchanged ideas with members of the gentry and aristocracy, like

Artis and Henderson's employer. See Anne Secord, "Corresponding Interests: Artisans and Gentlemen in Nineteenth-Century Lancashire," *History of Science* 27 (1994): 397–408, and "Science in the Pub," 284–88; R. J. Cleevely, "The Sowerbys and Their Publications in the Light of Manuscript Material in the British Museum (Natural History)," *Journal of the Society for the Bibliography of Natural History* 7 (1976): 362; and Noah Heringman, "'Stones so wonderous Cheap,'" *Studies in Romanticism* 37 (1998): 44–47, and the Introduction to this volume.

47. *NHPW* 347–49.

48. John Clare, *Early Poems of John Clare*, ed. Eric Robinson and David Powell, 2 vols. (Oxford: Clarendon, 1989), 1:211.

49. *John Clare*, ed. Eric Robinson and David Powell (Oxford: Oxford UP, 1984), 57–58, ll. 209–16.

50. Clare's Crow Flower is more typically called Corn Marigold or *Chrysanthemum segetum*.

51. For *Orchis morio*, see Sowerby, *English Botany*, 29:2059; for *Orchis mascula*, see Sowerby, *English Botany*, 9:631. V. S. Summerhayes discusses the natural hybridization between these two species in *Wild Orchids of Britain* (London: Collins, 1951), 112.

52. See William Curtis, *Botanical Magazine*, 42 vols. (London: W. Curtis, 1787–1817); Francis Bauer, *Illustrations of Orchidaceous Plants* (London, 1830–1838); James Bateman, *The Orchidacae of Mexico and Guatemala* (London: J. Ridgway, 1837–1843).

53. John Lindley and Thomas Moore, *Treasury of Botany*, 2 vols. (London: Longmans, Green, 1889), 2:820–21.

54. *Ladies' Botany* (London: Ridgway, [n.d.]), 229.

55. Phillip Cribbe, "Lindley's Life-Long Love Affair with Orchids," *John Lindley: 1799–1865*, ed. William T. Stearn (Woodbridge, Suffolk: Antique Collectors' Club, 1999), 107–42.

56. Lindley and Moore, *Treasury of Botany*, 2:820–21.

57. See, for example, the Orchis species discussed in *English Botany*, vols. 2 (1793), 5 (1796), 6 (1797), and 9 (1799).

58. *Cottage Tales*, ed. David Powell, P. M. S. Dawson, and Eric Robinson (Manchester: Carcanet, 1993), 145.

59. *Table Talk*, in *The Collected Works of Samuel Taylor Coleridge*, ed. Carl Woodring, 2 vols. (Princeton: Princeton UP, 1990), 1:342.

60. For a trenchant summary of recent shifts in the disciplinary assumptions of ethnography, see Marc Augé, *A Sense for the Other: The Timeliness and Relevance of Anthropology*, trans. Amy Jacobs (Stanford: Stanford UP, 1998), xii–xvi, 65–99.

61. A. L. de Jussieu, *Genera Plantarum*, intro. Frans A. Stafleu (1789; Weinheim: J. Cramer, 1964), xxiv, xxxvi, lxxii, 6–8; Atran, *Cognitive Foundations*, 188–89.

62. Augé, *A Sense for the Other*, 34.

CHAPTER NINE

Taxonomical Cures

The Politics of Natural History and Herbalist Medicine in Elizabeth Gaskell's Mary Barton

AMY MAE KING

> In the factory workers walking out in spring into Green Heys Fields; in Alice Wilson, remembering in her cellar the ling-gatherings for besoms in the native village that she will never again see; in Job Legh, intent on his impaled insects—these early chapters embody the characteristic response of a generation to the new and crushing experience of industrialism.
>
> —Raymond Williams

On the very night in Elizabeth Gaskell's 1848 *Mary Barton, A Tale of Manchester Life* that the eponymous heroine's mother will die in childbirth, Alice Wilson is invited to an impromptu supper at the Bartons' home; the younger Mary Barton runs to her cellar apartment and finds old Alice, like them just come in from a Sunday spent in Green Heys Fields just outside Manchester. Unlike the Wilsons and the Bartons, though, Alice has not been taking a social walk:

> she had been out all day in the fields, gathering wild herbs for drinks and medicine, for in addition to her invaluable qualities as a sick nurse and her worldly occupations as a washerwoman, she added a considerable knowledge of hedge and field simples; and on fine days, when no more profitable occupation offered itself, she used to ramble off into the lanes and meadows as far as her legs could carry her. This evening she had returned loaded with nettles, and her first object was to light a candle and see to hang them up in bunches in every available place in her cellar-room.[1]

Her nettles, intended as she tells Mary for "spring drink," are not the sentimental plucking of a Sunday walk; their stinging properties belie that, as does the more general description of her humble cellar room,

> oddly festooned with all manner of hedge-row, ditch, and field-plants which we are accustomed to call valueless, which have a powerful effect either for good or for evil, and are consequently much used among the poor. The room was strewed, hung, and darkened with these bunches, which emitted no fragrant odour in their process of drying. In one corner was a sort of broad hanging shelf . . . where some old hoards of Alice's were kept. (*MB* 17)

Alice Wilson's gatherings suggest an alternative medical practice, one whose exact relation to vernacularized medicine and natural history alike has remained beyond the reach of critical clarity. In our accounts of eighteenth- and nineteenth-century medicine, we have begun to recognize and describe alternative medical practices, including demotic or folk traditions as well as spheres such as quackery previously thought outside any traditional understanding of the history of medicine.[2] That this has yet to be registered in relation to Gaskell's mid-century social fictions—no stranger, as the characters of Dr. Gibson and Roger Hamley of *Wives and Daughters* (1866) suggest, to the figure of the professional scientist or doctor—is evident in the failure to recognize Alice Wilson's vocation beyond that of her status as a washerwoman.[3] In our persistence in apprehending characters through an exaggerated emphasis on professional forms of scientific medicine, we not only risk misunderstanding the social placement of a single character but also the discourse through which Gaskell's novelistic representation of "social ills" is best achieved.[4] My purpose here is to sketch an account of the lingering vernacular medical tradition in the nineteenth century—specifically, the vernacular medicine of the herbalist—and to suggest two important facets of this tradition: its deep epistemological links with natural history and the importance of this medical epistemology to novelistic representations of social ills. In *Mary Barton*, the physical ills that Alice's herbalist medicine intends to cure are extended to the "remedies" for social ills by one important bridge: a cure for what I will call the "ills of perception" that the taxonomical basis of natural history implicitly suggests.

Insofar as *Mary Barton* has been understood as one of the industrial novel's chief prototypes, the choice to look to the epistemologies of natural history in order to account for the text's political energies may seem puzzling. It is perhaps less odd when one turns briefly to Gaskell's life, for her biography suggests that she was conversant with natural history topics and the emergent scientific questions of the early Victorian era. Jenny Uglow states that in 1830 Gaskell went to stay with her relative the Reverend William Turner, a pioneering figure in linking local society with larger scientific movements of the day. Just during her immediate stay in his home, he gave three lectures at the Society for the Promotion of Natural History and the

Literarary and Philosophical Society: "The Vegetable Kingdom," "Mineralogy and Geology," and "Optics and Astronomy."[5] If Manchester had been a scientific backwater, during the 1820s and 1830s it inaugurated five separate scientific societies and even hosted the annual meeting of the British Association in 1842.

Two versions of amateur natural history are represented in *Mary Barton*: Job Legh's studies of botany and entomology place him squarely within the tradition of the nineteenth-century natural historian, while Alice Wilson's knowledge of medicinal plants evokes an older tradition of the village herbalist or wisewoman. Although *Mary Barton* most obviously is a sympathetic documentation of the sufferings of the industrial poor that Gaskell herself witnessed as the wife of a Manchester minister, it also turns to the occupations and pleasures of certain of those very same people:

> weavers, common hand-loom weavers, who throw the shuttle with unceasing sound, though Newton's "Principia" lies open on the loom . . . there are botanists among them, equally familiar with either the Linnaean or the Natural system, who know the name and habitat of every plant within a day's walk . . . there are entomologists, who may be seen with a rude looking net, ready to catch any winged insect. (*MB* 37–38)

Gaskell's attempt to capture a certain documentary reality—what Raymond Williams called her "intensity of effort to record, in its own terms, the feel of everyday life in the working-class homes"—included the realities of amateur natural history and vernacular medicine among the working-class industrial poor.[6] The elaborate references to working-class scientific pursuits gesture to one of the ways in which *Mary Barton*, although generically eclectic as Catherine Gallagher has shown, strives to be a documentary record of and argument for an enlightened proletariat.[7] The fact that the novel opens outside the city, in a space characterized by natural history collecting—the source of Alice's herbal medicine—and situates its working-class denizens amid narrative energies that verge on the taxonomic, suggests the importance of natural history to this industrial novel.

In *Mary Barton*, Gaskell invites her readers to see the rural not as a flight from contemporary urban–industrial reality but as a literal cure for its worse ills, the source of the very sort of medical and educational relief for which reform agitates.[8] Although Alice Wilson is nostalgic for the rural village she left as a child, her herbalist work, like the naturalist studies of Job Legh, is resolutely work for the present; her knowledge of herbal "simples" is rural in origin but urban in practice: she gathers nettles for "spring drink," and dries herbs and other plants, to revive the health of her neighbors and prepare for future illness. Gaskell's remedy for the social ills she depicts, like Alice's remedies, is present-directed and firmly unnostalgic: it is not a position that urges a return to the country, a position underscored by the desperation of the replacement workers ("knob-sticks") who come from the outlying countryside.[9] The text's representation of herbalist medicine and the study of nature is not nostalgic but progressive, for they suggest a model for reform in a novel that has been described as dramatizing, but not offering remedies for, social ills.

However, the true importance of herbalist medicine to the text is not simply thematic but structural; the city's outer reaches not only contain the symbols of proposed reforms but also, more significantly to Gaskell's narrative purpose, suggest a reform of perception. That is, Gaskell uses the taxonomical logic of natural history to propose a more humane way of seeing the working class, not as another species or as types of a larger order but as individual specimens whose ills can be ameliorated. Gaskell's most sustained and coherent political call in *Mary Barton* is for the two classes to *see* each other and classify themselves as like species. Chartism, which is the novel's avowed politics, is invoked but not elaborated on, while the text oddly has sustained and detailed references to natural history. If natural history does not seem political, it also is in no way oppositional to the political energies of working-class movements; in fact, as Friedrich Engels wrote in his 1844 *Condition of the Working Class in England*, working-class institutes run by trade unions and socialists sponsored lectures, the most popular of which were on "economics and on scientific and aesthetic topics."[10] That natural history was a site of connection between political and medical heterodoxy has also been well established: for instance, Isaiah Coffin, the leader in England of a working-class medical self-help movement centering on herbalism, called for the establishment of local botanical societies to provide working-class access to practical plant knowledge.[11] Moreover, the amateur natural historian Job Legh is not apolitical; aside from being John Barton's most faithful listener, he is a symbol of the kind of working-class enlightenment that reformers supported.

Whatever sympathy Gaskell brought to the documentation of everyday life for the industrial poor of the 1840s, she also brought a strong opinion about the source of their problems: a misperception on the part of the industrialists about the nature of the working class and the cause of their social ills. As such, Gaskell's challenge was to balance the representation of working-class suffering with her belief that poverty was not natural but a state engendered by industrial practices.[12] Through Alice Wilson's herbalist remedies and Job Legh's naturalist studies—two seemingly discordant, but nevertheless persistently evoked, practices in this industrial novel—*Mary Barton* ultimately suggests a cure for the "ills of perception" that inform the suffering and strife in the novel.

Herbal Medicine in *Mary Barton*

Alice Wilson's herbal knowledge in *Mary Barton* points us toward the complex and long history of vernacular medicine in England. Clearly, Alice's knowledge does not stem from a professional, or perhaps even literate, tradition, but rather a tradition of proverbial lore. Poor enough that she drinks herbal rather than black tea and lives in a cellar, Alice may not have been literate and most likely would not have owned an herbal. Her knowledge is folk knowledge, popular, rural, and most likely eighteenth-century in origin and nature; Alice was raised in Cumberland, we are told, but emigrated to Manchester as a young woman. Her social position as a sick nurse and herbalist, as well as washerwoman, gestures back to a much older and, by and large, fading vernacular medical tradition in England. In early modern England, for instance, aside from medical professionals of various sorts, there were, as Roy Porter has shown, informal healers: people who "practiced healing without a view to reward, but out of motives of neighborliness, paternalism, good housekeeping, religion or simple self help. Every village had its 'nurses' and 'wise women' well versed in herb-lore and in secret brews and potions."[13] By the Georgian and early Victorian periods, with the increasing dominance of regular medicine, the tradition of the "wise woman" was fading, with Victorian collectors of folk medicine believing they were preserving a dying art.[14]

And yet it would be a mistake to create a simple opposition between a fading amateur tradition and a rising professionalized medicine; Alice's knowledge represents not simply a disappearing art of healing but an established tradition of vernacular medicine that had staying power throughout the eighteenth century and into the nineteenth. Although Alice's knowledge seems predominantly proverbial and oral, the printed record of that knowledge is retained in a fashion in the herbals that remain. The most influential early English herbal was the early Latin herbal of Apuleius Barbarus (or Pseudo-Apuleius); the chief source of botanical knowledge in Europe during the so-called Dark Ages, it was probably translated into Anglo-Saxon about A.D. 1000 and revived as a source with its mid-nineteenth century translation in the compilation known as *Leechdom's Wortcuning and Starcraft of Early England.* Other important English herbals, such as the *Grete Herball* (printed in 1526), *Bancke's Herball* (1525), Gerard's *Herball* (1633), and *Hortus Sanitatis* (1485), formalize folk knowledge; they picture the herb through a woodcut, describe the plant, list its popular and botanical names, and describe its "virtues"—that is, its medicinal uses. The healing virtues of an herb might stand alongside suggestions for how to poison a pest or flavor a soup.

The diffusion of medicine to the public had long taken two channels: word of mouth and printed books purveying medical advice. Thus, it is not entirely clear if Alice's herbal knowledge comes strictly from proverbial lore or if her knowledge stems in part from printed vernacular medical sources. That is, although her herbal knowledge has links to the practical advice of the medieval and early modern herbals, her self-help model also suggests a popular medical culture that thrived throughout the eighteenth century and into the nineteenth, despite the rise of professional and paraprofessional medicine among the middle classes beginning in the eighteenth century.[15] Like the residual Anglo-Saxon in John Barton's northern, working-class dialect, Alice's herbal knowledge is both ancient and thoroughly contemporary. That mixed status—a domain of medical knowledge that operated both within and without formal texts and remained viably modern for all of its long heritage—is attested by publication histories: John Culpeper's *Herbal* from the mid-seventeenth century on had assumed "exemplary status as a household name," while William Buchan's *Domestic Medicine* (1769) was reprinted well into the nineteenth century.[16] Domestic manuals—medical lore books, such as Eliza Smith's *The Compleat Housewife* (1729), that contained an assortment of household hints, recipes, and medical cures—were another way in which medical knowledge accumulated.[17] Nevertheless, as Porter has shown, although much "lay therapeutic lore was put down on paper . . . it circulated and snowballed in many other ways."[18]

John Wesley's 1747 *Primitive Physic, or an Easy and Natural Method of Curing Most Diseases* played a large role in the diffusion of medical knowledge, particularly within his chosen audience, the working poor. Wesley knew that going to a doctor or buying prepared medicine was financially difficult, so his book offers traditional remedies for disease:

> a mean hand has made here some little attempt towards a plain and easy way of curing most diseases. . . . Who would not wish to have a Physician always in his house, and one that attends without fee or reward? . . . Is it enquired, but are there not books already, on every part of the art of medicine? . . . [T]hey are too dear for poor men to buy, and too hard for plain men to understand. . . . I have not seen one yet, either in our own or any other tongue, which contains only safe, and cheap, and easy medicines.[19]

Wesley's *Primitive Physic* went through many editions, inspiring a formal revival of interest in herbal medicine just prior to the era when it would be most endangered; as the historian Mary Chamberlain has claimed, Wesley's tract "enabled that tradition to continue when the social and practical supports had been destroyed by emigration from the villages to the city."[20] Herbalism, then, had roots in preindustrial England, but it also, tellingly, had a nineteenth-century history that included the new medical cosmology known as medical botany. "Thomsonianism," after the American herbalist Samuel Thomson, or Coffinism, after Thomson's English follower Isaiah Coffin, rejected orthodox medicine in favor of natural herbal remedies; their following was primarily in the artisan and petit bourgeois social zones, but their rhetoric was populist and sympathetic to heterodox political movements such as the Owenites and Chartists. Alice's herbal knowledge, however, seems too proverbial—as the result of informal exchange networks—to be derived from these movements and might be traced with more accuracy back to her rural background. Notwithstanding Alice's folk-derived knowledge, the political tendencies of a newly publishable herbalism remain suggestive and well within the range of Gaskell's own social sphere.

Alice's herbalist remedies are part, therefore, of a more general community of medicine that would have included the informal and generally unremunerated work of midwifery and laying out; Alice may have performed each of these roles, as well as making medicines, despite nominally being employed as a washerwoman. George Wilson brags that "there's not a child ill within the street, but Alice goes to offer to sit up, and does sit up too, though may be she's to be at her work by six the next morning" (*MB* 12). Later, Alice's nephew strikes a similar note: "I used to be wakened by the neighbours knocking her up; this one was ill, and that body's child was restless; and for as tired as ever she might be, she would be up and dressed in a twinkling. . . . How pleased I used to be when she would take me into the fields with her to gather herbs . . . she knew such a deal about plants and birds, and their ways" (*MB* 194). Among the working classes, access to doctors was limited throughout the nineteenth century; folk medicine, as well as informal aids for childbirth and illness, filled the gap, as Alice's medicine and the several allusions to neighbors functioning as nurses suggest. When John Barton goes to the apothecary on behalf of the starving and fever-ridden Ben Davenport, he has to pawn his valuables; the prosperous-looking shop darkens Barton's mood, where he is given "sweet spirits of nitre," a medication Gaskell informs us is useless, but which Barton took with "comfortable faith . . . for men of his class believe that every description is equally efficacious" (*MB* 63–64). Engels had sharper criticism for plyers of patent medicines, calling them "charlatans" and their medicines "quack remedies."[21] In contrast, the nettles that Alice gathers, or the "meadow-sweet" that she sought out "to make tea for Jane's cough," are clearly medicinal references to plants; by contrast to her scathing criticism of the opium-dispensing apothecaries, the narrator never criticizes or condescends about Alice's simples.[22] Alice's medicines represent an alternative medical culture that reaches back into a preindustrial past to cure the ills that literally surround her. More crucially, perhaps, her medical practice comes out of the herbal tradition, which marries medical knowledge with classificatory botanical knowledge; it is a combination that points the way to the other body of naturalist knowledge in *Mary Barton*, Job Legh's natural history.

Manchester's Natural Historian

If Alice's herbalism reaches back to an early modern paradigm of naturalizing, *Mary Barton* presents us with a modern naturalist, Job Legh, whose natural history enthusiasm finds a vocabulary and structure in a slightly more contemporary source: the classificatory schemes of eighteenth- and nineteenth-century taxonomy. Nor is Job alone in these pursuits: we might recall the weavers who read Newton and know the Linnaean system, figures who evoke a working-class familiarity with natural history. Gaskell's narrative method—primarily that of the sympathetic observer, or documenter, of working-class life—is not unlike the method of natural history: both emphasize, to the point of bias, visuality because both privilege vision over the other senses as the mode for conveying description and character-zation. It is a bias that gives sight, as Foucault demonstrates, an "almost exclusive privilege" in natural history classification, and makes it the dominant mode of representation in the novel form: the blind man can "perfectly well be a geometrician, but he cannot be a naturalist."[23]

The epistemological links between novel writing and the practice of natural history are concretized in the suggestive mapping of taxonomical language and situation onto the human sphere. In the passage that immediately follows the narrative revelation that John Barton had become "a Chartist, a Communist, all that is commonly called wild and visionary," we learn of his "rough Lancashire eloquence," his particular talent for "method and arrangement," and most important, the fact that "he was actuated by no selfish motives; that *his class, his order*, was what he stood by, not the rights of his own paltry self" (*MB* 170; emphasis added). His class, his order: these are words that separately have sociopolitical significations (as perhaps they are at first understood here) but which in combination also suggest the taxonomical organization of the natural world: kingdom, phylum, class, order, family, genus, species. *Mary Barton* opens with a scene that seems to situate humans within a similar taxonomical logic; walking in the fields outside the city, we learn that the landscape is populated with types of people—specifically, toddlers, girls, boys, whispering lovers, husbands, and wives—all grouped by Gaskell into a totalizing mass: "the manufacturing population" (*MB* 6–7). As the narrative narrows its focus from the group to an individual, it focuses on what it calls "a thorough specimen of a Manchester man." A "specimen," then, of the order known as the manufacturing population, walking amid a new spring day, surrounded by specimens of nature that the narrative is just as eager to detail, such as the "young green leaves, which almost visibly fluttered into life" (*MB* 6).

Job Legh never appears in the novel without reference to his natural history avocation; repeatedly, his appearance brings with it a sustained discussion or at the very least a minor reference. In one instance, his speech is paired with this description—"for Job Legh directly put down some insect, which he was impaling on a cork-pin, and exclaimed"—while in another, we learn that he is absent from the house because he "has been out moss-hunting" (*MB* 355, 45). In another instance, Margaret explains why she continues to sew, despite the threat of blindness, by bringing up natural history: "grandfather takes a day here, and a day there, for botanising or going after insects, and he'll think little enough of four or five shilling for a shilling: dear grandfather! And I'm so loath to think he should be stinted of what gives him such pleasure" (*MB* 48). When Margaret goes on tour to sing, Job Legh shuts the house for—what else?—natural history: "her grandfather, too, had seen this to be a good time for going on his expeditions in search of specimens" (*MB* 141). In failing to write from Liverpool, Job is excused because "writing was to him little more than an auxiliary to natural history; a way of ticketing specimens"

(*MB* 340). The novel, of course, ends with Job, who is contemplating coming to Canada not to see them, but to "pick up some specimens of Canadian insects . . . all the compliments . . . to the earwigs" (*MB* 393). This odd habit of tagging Job Legh's appearances in the text with references to natural history keeps natural history present to the reader; in light of the sustained quality of the references, the more extended discussions of natural history topics in the novel seem to be more than mere curiosities.

One such discussion, on the topic of a flying fish, introduces taxonomical language into the text: Job knows the flying fish is "the Exocetus, one of the Malacopeterygii Abdominales" (*MB* 153). On two occasions the difference between vernacular and taxonomical language is discussed; on this occasion, Alice's nephew Will says, "you're one o' them folks as never knows beasts unless they're called out o' their names. Put 'em in Sunday clothes, and you know 'em, but in their work-a-day English you never know nought about 'em" (*MB* 153). When Mary meets Job for the first time, she is "bewildered" by the "strange language," the "technical names which Job Legh pattered down on her ear" (*MB* 40). These overt references to Job's learnedness separate him from Alice Wilson, whose references to plants are in an untutored vernacular. In both cases, however, these naturalists do not represent a desire to return to a nostalgic past: the social ills produced by the industrial context are not meant to be remedied by a return to the country because Gaskell is unsentimental about rural poverty.

The fact that Job Legh and Alice Wilson thrive amidst a generally more bleak landscape is itself telling: Alice dies of old age, not of illness or starvation, and raises a foster son to robustness; in a time and place where "very few strong, well-built, healthy people are to be found among them," Will's physical health and beauty distinguish him.[24] Job, perhaps more astoundingly, is happy and enlightened; despite the absence of teeth, his eyes "gleamed with intelligence" as he sits amid a room lined with "rude wooden frames of impaled insects . . . [and] cabalistic books" (*MB* 40). Moreover, his education enables a practical competence that enables Jem's defense; Mary, who was overwhelmed by the word "alibi," thinks to consult Job because he is "gifted with the knowledge of hard words, for to her, all terms of law, or natural history, were alike many-syllabled mysteries" (*MB* 247). Gaskell goes so far as to link Job's ability to secure a lawyer with his knowledge of natural history; he knows a lawyer, a "Mr. Cheshire, who's rather given over to th' insect line . . . he and I have swopped specimens many's the time, when either of us had a duplicate" (*MB* 260).

Job's dedication to natural history is perhaps most evident in an extended story about a scorpion. He acquires the scorpion by walking to Liverpool in order to inquire of sailors if they had accidentally brought a specimen back from overseas. The story of the scorpion has usually been read in relation to the passage in the chapter entitled "Return of the Prodigal": "the people had thought the poverty of the preceding years hard to bear, and had found its yoke heavy; but this year added sorely to its weight. Former times had chastised them with whips, but this chastised them with scorpions" (*MB* 114). Gaskell here is deliberately echoing 1 Kings 12:11: "my father made your yoke heavy, and I will add to your yoke: my father chastised you with whips, but I will chastise you with scorpions." Elaine Jordan, in light of the context of the story in 1 Kings, understands the story of the torpid scorpion as a parable of the present labor situation; the scorpion is "apparently a random choice" except for its symbolic association with the biblical passage.[25]

I would argue that even if the scorpion symbolically invokes rebellious slaves, it also is situated within a tale of natural history collection, and as such suggests intellectual curiosity and equality. In this way the scorpion may very well invoke the biblical context of slavery pushed too far, but it also is a symbol of enlightenment. Job acquires the scorpion by walking to Liverpool in order to possibly acquire a specimen from a place he could never afford to go; as such, it is an act of pure intellectual curiosity. In neglecting the context of this resonant image, we neglect his avocation—the very thing he performed outside the context of remunerated labor; in neglecting natural history as such in order to emphasize the image of the scorpion as a biblical image of punishment and rebellion, we neglect the potency of the image of the enlightened working class. Here, the biblical allusion and the natural history context work together to make a similar political point: together they advocate for moderation and enlightenment for the working classes.

Cures for Perceptual Ills

What, then, might the implications of this natural history practice—as well as a similarly taxonomic herbalist medicine—be for an understanding of the social ills that my examples have heretofore seemingly skirted? Although Gaskell does not efface her knowledge of natural history in the way that she falsely disowns knowledge of "political economy" in the preface, the political implications of natural history are nevertheless never made explicit in the text. And yet natural history's methodological reliance on sight, and its emphasis on classification, perhaps suggest a more coherent basis for the novel's politics than John Barton's avowed, but incompletely realized, Chartism. Gaskell obliquely proposes that the deterioration of the relationship between what Engels would describe as "the industrial proletariat" and the "capitalists" is caused by a *perceptual* error: the class system that enforces such a wide economic divide between the working-class denizens of Manchester and their capitalist bosses threatens the correct perception that each man, regardless of social position, is of the same species.

Catherine Gallagher has noted that "all versions of John Barton's life thus become irrelevant to the novel's concluding and redeeming action: Carson's forgiveness, which is a foretaste of the Christian spirit that the narrator assures us will allow Carson to effect industrial social change."[26] Carson's forgiveness is Christian in its structure, but it is based in sympathetic identification achieved by a change in perception. As we will see, the penitence of John Barton is matched, at the novel's climactic meeting of the two, by Mr. Carson's renunciation of violence and commitment to remedying his workers' lot. A humanitarian's fantasy, perhaps, but also a serious challenge to the very social distinctions that drive the social ills that the novel portrays. If Gaskell's politics reside most comfortably in a kind of familiar humanitarianism, it is a humanitarianism not restricted to sympathy or pity for the industrial poor; instead, Gaskell's humanitarianism more radically places the responsibility on the industrialist to reorder his categorization of his workers as less human than himself.

The fact that *Mary Barton* also contains an herbal medical tradition is telling in light of the fact that it is a novel about social ills ending with a projected "remedy." The social ills that are most clearly articulated in the novel are environmental and domestic; following John Barton's death, Jem talks with Mr. Carson about the working class's need to know that a remedy was being sought; "remedy" is then repeated twice to describe the amelioration of social suffering. Engels too uses the image of disease to discuss the social situation of the working class in England: "the social disorder from which England is suffering is running the same course as a disease which attacks human beings."[27] Engels of course is unlike Gaskell in that he would hasten the disease in order to promote socialist revolution, while Gaskell has perhaps been justifiably accused of dramatizing the "fear of violence" among the upper and middle classes at the time.[28] What, then, is Gaskell's proposed cure, and how is it linked to natural history and herbal medicine? The herbalist tradition suggests a cure based in a self-reliant folk knowledge, a cure that would put to use a knowledge that already exists. The methodology of natural history requires that one visually examine a species that one *sees* in order to make an analytical judgment. The remedy that is being proposed through the cognate fields of herbalist medicine and natural history, then, is a combination of self-reliance and change in perception between the different social classes.

In this light, one can see that the tragic arc of *Mary Barton* is structured around the loss and regaining of the perception of the other as human. The violent explosion that leads to Harry Carson's murder is incited by a caricature, the drawing that makes fun of the trade-union men's appearance by likening them to Falstaff's clowns: he "had drawn an admirable caricature of them—lank, ragged, dispirited, and famine-stricken. Underneath he wrote a hasty quotation from the fat knight's well-known speech in Henry IV" (*MB* 184).[29] That the men cannot laugh in light of their hunger is unsurprising; it marks the end of sympathetic imagination on the part of the workers because they plot the murder on the back of the very piece of paper that had caricatured their hunger. The workers' shift in sympathetic imagination occurs when the caricature tangibly reveals the owners' perception that the working-class men are of a different order, a different species of man. Of course, the reader had been privy to this already; the owners call them brutes and animals, and most damagingly nonhuman: "I for one won't yield one farthing to the cruel brutes; they're more like wild beasts than human beings"(*MB* 182). Their lack of sympathetic identification, however, is literalized in the caricature; their failure to correctly perceive them as human proves to be the final indignity. In not empathizing with their starvation and caricaturing their plight, the owners upset the fragile sympathy upon which the civil relationship between the two classes of men had been based.[30] The caricature turns violent what had been a linguistic tendency toward the breakdown in sympathetic imagination; that is, the linguistic tendency, dramatized in the novel's first pages, to divide the world into "employers and employed": "the differences between the employers and employed . . . [are] an eternal subject for agitation in the manufacturing districts, which, however it may be lulled for a time, is sure to break forth again with fresh violence at any depression of trade" (*MB* 23).

The industrialists' failure is a failure of perception and classification; their breach is in misclassifying their workers as brutes, hence setting in motion the cycle of violence that the novel then depicts. Job Legh's natural history avocation is thus not only a detail about working-class life but a structural model for reform: it offers a cure for the ills of perception. In seeing the affinity between the working-class man and himself, the industrialist rights the wrongs of misclassification that had set in motion the violence. Of course, with that perceptual change comes a return of sympathetic imagination—for the elder Carson, the incitement to sympathy occurs in the house of John Barton, where he has been summoned to hear the man's confession. In thinking about it in his study afterward, it is the visual impact of the distraught man's poverty that elicits his empathy:

> In the days of his childhood and youth, Mr. Carson had been accustomed to poverty; but it was honest, decent poverty; not the grinding squalid misery he had remarked in every part of John Barton's house, and which contrasted strangely with the pompous sumptuousness of the room in which he now sat. Unaccustomed wonder filled his mind at the reflection of the different lots of the brethren of mankind. (*MB* 370)

The promise of a confession to his son's murder had brought Carson to the Bartons' house, a face-to-face confrontation with the poverty the trade unions had described but which he had been unable to perceive before literally *seeing* it here. The return of a Christianized, humanitarian vocabulary—the "brethren of mankind"—marks the return of sympathy, a sympathy that had begun at Barton's but which he had initially rejected in order to pursue vengeance. Barton's confession revolves around a similar transformation, what the text calls the return of "sympathy for suffering, formerly so prevalent a feeling with him, again filled John Barton's heart." The change in perception is articulated by the narrative, which sees the return of sympathy as the cessation of false difference: "the mourner before him was no longer the employer, a being of another race, eternally placed in an antagonistic attitude . . . no longer the enemy, the oppressor, but a very poor and desolate old man" (*MB* 366). Although the scene climaxes with a reference to the Old Testament—"forgive us our trespasses"—it is the mutual shift in perception that gets the two men to the religious proverb. The linguistic tendency that reveals the threat to perception is overcome; in not seeing him as a "being of another race" but as a "man," Barton initiates the political cure that Gaskell is implicitly advocating; in seeing him anew, "no longer the employer . . . the enemy . . . the oppressor," Barton's perception is reformed, which in turn models the reform required for the cure of social ills.

Could the nettles that Alice had gathered on that spring Sunday have been intended to aid in the approaching childbirth of John Barton's wife, with its possible, even probable, complications? Pliny, a source for English herbals, suggests nettle juice for "prolapsus of the uterus," a specific uterine complaint that the English herbalist tradition both picks up on and broadens. The breadth of "virtues" in plants, however, is so great as to defy simple analogical connections between medical and literary text. In many ways, Pliny's recording of folk wisdom about nettles registers the irony of Mrs. Barton's death more completely than the suggestive connection between female illness and nettles; he records that in the spring nettles are eaten with the "devout belief that it will keep diseases away throughout the whole year."[31] That we can trace John Barton's despair, and subsequent radical political thoughts and actions, to the death of his wife is all the more suggestive about the import of herbal medicine and natural history to the politics of *Mary Barton*. If the irony of Alice's nettles arriving too late is less present to us than the arrival of a doctor only after Mrs. Barton's death, it is only because we have forgotten the prevalence of nineteenth-century herbalist medicine for the working classes. In the persistent evocation of natural history and herbal medicine, the techniques of reformist realism are broadened. Gaskell, in a sense, proposes a taxonomical cure: a way of *seeing* the working class, not as brutes, or just as "the manufacturing population," but as individuals whose ills should be ameliorated.

Notes

The epigraph to this chapter is from *Culture and Society 1780–1950* (London: Chatto and Windus, 1958), 88.

1. *Mary Barton, A Tale of Manchester Life*, ed. Macdonald Daly (New York: Penguin, 1996), 16. Cited parenthetically in the text hereafter as *MB*.

2. See, for instance, Roy Porter, *Health for Sale: Quackery in England, 1660–1850* (Manchester: Manchester UP, 1989) and *In Sickness and in Health: The British Experience,*

1650–1850 (London: Fourth Estate, 1989); Anne Digby, *Making a Medical Living: Doctors and Patients in the English Market for Medicine, 1720–1911* (Cambridge: Cambridge UP, 1994); G. Gisse, R. L. Numbers, and J. W. Leavitt, eds., *Medicine without Doctors* (New York: Science History Publications, 1977); W. F. Bynum and Roy Porter, eds., *Medical Fringe and Medical Orthodoxy, 1750–1850* (London: Croom Helm, 1987).

3. For an interesting account of the medical and scientific contexts in *Wives and Daughters*, see Deirdre D'Albertis, *Dissembling Fictions: Elizabeth Gaskell and the Victorian Social Text* (New York: St. Martin's, 1997).

4. Narratives of professionalization in the Victorian period are, in fact, often based on the development of the medical profession. As a result, literary–critical interest in Victorian medicine has emphasized the growth of its contemporary, professional contours—perhaps so much so that we have failed to recognize a medical scene in *Mary Barton* that lies outside that professional context. For excellent accounts about the discourse of medical professionalization, see Penelope Corfield, *Power and the Professions in Britain, 1700–1850* (London: Routledge, 1995); Magali Sarfatti Larson, *The Rise of Professionalism: A Sociological Analysis* (Berkeley and Los Angeles: U of California P, 1977).

5. *Elizabeth Gaskell: A Habit of Stories* (Boston: Faber and Faber, 1993), 213.

6. Williams, *Culture and Society*, 87.

7. *The Industrial Reformation of English Fiction: Social Discourse and Narrative Form, 1832–1867* (Chicago: U of Chicago P, 1985).

8. As such I tend to disagree with those who see the opening scene of the novel as a nostalgic representation of what has been lost because of industrialization—what Coral Lansbury calls a "lost arcadia . . . a landscape of a lost world that can only be visited on holidays and in the dreams of nostalgia." Lansbury, *Elizabeth Gaskell: The Novel of Social Crisis* (New York: Harper and Row, 1975), 25.

9. Even when Alice becomes senile after her stroke, her mental retreat to the countryside is less a nostalgic evocation of a better time and place than a continuation of her practical approach to nature. For instance, one of the places her dementia returns her to is the fields she went to with her sister; the "heather and ling for besoms" that she recalls searching for is a reference to the folk knowledge she embodies, for heather and ling (similar broomlike plants) were gathered to make brooms, or "besoms." Neither are the scenes of her returned childhood hazy or romantic; they are specific in their evocation of the natural world: "the bees are turning homeward for th'last time, and we've a terrible long bit to go yet. See! Here's a linnet's nest in this gorse-bush. Th'hen bird is on it. Look at her bright eyes, she won't stir. . . . Won't mother be pleased with the bonny lot of heather we've got" (*MB* 215).

10. *Condition of the Working Class in England*, trans. W. O. Henderson and W. H. Chaloner (Stanford: Stanford UP, 1968), 272. Engels makes a distinction between the mechanics institutes, which he considers middle-class propaganda machines, and working-class institutes run by trade unions. He writes that "the middle-classes hope also that by fostering such studies they will stimulate the inventive powers of the workers to the eventual profit of the bourgeoisie" (271). He states that the socialists and trade unions do a far better job at educating workers, some of whom he admires for a very high degree of natural history learning: "I have sometimes come across workers, with their fustian jackets falling apart,

who are better informed on geology, astronomy, and other matters, than many an educated member of the middle classes in Germany" (272). One way in which Gaskell's novel seems to conform to the distinction that Engels would draw between the Mechanics Institutes and educational programs sponsored by trade unions is the contrast between Jem Wilson's and Job Legh's learning. Jem makes an "invention for doing away wi' the crank, or somewhat. His master's bought it from him, and ta'en out a patent" (*MB* 142). In inventing a new industrial tool, Jem represents in no small measure the middle-class fantasy that Engels describes with such derision; his knowledge, that is, can be coopted by the bourgeoisie. Job Legh, on the other hand, pursues knowledge without any productive or economic value and, hence, more fully represents the kind of educational enlightenment that Engels valued.

11. J. F. C. Harrison, "Early Victorian Radicals and the Medical Fringe," *Medical Fringe and Medical Orthodoxy: 1750–1850,* eds. W. F. Bynum and Roy Porter (London: Croom Helm, 1987), 198–215.

12. Hilary Schor's work on *Mary Barton* makes a compelling case for the novel's success as political critique. See Schor, *Scheherezade in the Marketplace: Elizabeth Gaskell and the Victorian Novel* (New York: Oxford UP, 1992).

13. *Disease, Medicine, and Society in England, 1550–1860* (London: Macmillan, 1987), 21.

14. Ibid., 43–44.

15. See Mary Chamberlain and Ruth Richardson, "Life and Death," *Oral History: Journal of the Oral History Society* 11.1 (1983): 41; Porter, *Health for Sale.*

16. See Roy Porter, "Introduction," *The Popularization of Medicine: 1650–1850,* ed. Roy Porter (London: Routledge, 1992), 2.

17. Philip Wilson, "Acquiring Surgical Know-How: Occupational and Lay Instruction in Early Eighteenth-Century London," *The Popularization of Medicine: 1650–1850,* 47–57.

18. *In Sickness and in Health,* 268.

19. *Primitive Physic: or an Easy and Natural Method of Curing Most Diseases,* ed. A. Wesley Hill (London: Epworth, 1960), 27–28.

20. Ibid., 32.

21. Engels writes about the propensity of the working class to be swindled by medical charlatans. He discusses the Manchester Infirmary, an institution that he praises but which in 1832 could only treat 22,000 patients, a number which could not nearly meet the needs of the population. However, the relationship between the Manchester that Elizabeth Gaskell saw and the Manchester of Engels's narrative is tenuous at best. Susan Zlotnick suggests that "one might forget that the two accounts claim to represent the same city at the same historical moment." Zlotnick, *Women, Writing, and the Industrial Revolution* (Baltimore: Johns Hopkins UP, 1998), 78.

22. Their utility is explicitly contrasted with the Carsons' discussion of plants, whose interest is strictly aesthetic. The youngest Carson daughter begs her father for a "new rose," a species that costs an astounding half-guinea; she calls it "one of her necessaries," and pouting, says that she does not consider dandelions and peonies flowers (*MB* 69–70). Tellingly, the flowers that she rejects in favor of the new hybrid rose are plants that can be used;

dandelion greens may be eaten, and the seeds of peonies induced labor, while its roots were thought to cure palsy and lunacy.

23. Michel Foucault, *The Order of Things: An Archeology of the Human Sciences* (New York: Vintage, 1994), 133.

24. Engels, *Condition*, 118. The statistics that Engels cites are stark: 54 percent of the working poor's children did not survive.

25. "Spectres and Scorpions: Allusion and Confusion in *Mary Barton*," *Literature and History* 7 (1981): 59.

26. *Industrial Reformation*, 87.

27. *Condition*, 139.

28. Williams, *Culture and Society*, 90.

29. Gallagher points out that the younger Mr. Carson is killed for turning a "worker's delegation into . . . Shakespearean clowns." Gallagher's reading of Gaskell's use of Shakespeare with the caricature speaks to Gaskell's social critique of literature. She writes, "Carson's destructive use of Shakespeare reminds Gaskell's readers that although they have the best precedents for laughing at rags and tatters, they must now free themselves from the conventional association between 'low' characters and comedy" (*Industrial Reformation* 69–70).

30. Raymond Williams writes about sympathetic imagination, but he does so in relation to the writing act rather than within the bounds of the plot. That is, his reading of *Mary Barton* discusses Gaskell's sympathetic imagination and the threat made to it by the "fear of violence"; Williams acknowledges her "flow of sympathy" and evaluates the novel as a "largely successful attempt at imaginative identification," but suggests that the fear of violence explains her dramatization of the murder (*Culture and Society* 90–91).

31. Plinius Secundus, C., *Natural History, Vol. VI*, trans. W. H. S. Jones (Cambridge: Harvard UP, 1938), 229, 315.

About the Contributors

ALAN BEWELL is professor of English at the University of Toronto. Focusing on the relationship between literature, science, and medicine, he has published *Wordsworth and the Enlightenment* (1989) and *Romanticism and Colonial Disease* (1999), as well as essays on topics such as botany and literature, eighteenth-century obstetric theory, and Joseph Banks. He has also edited *Medicine and the West Indian Slave Trade* (1999). He is currently working on a monograph on Romanticism and natural history.

RACHEL CRAWFORD is associate professor of English at the University of San Francisco. She has published articles on romantic and georgic poetry in *Studies in Romanticism, Romanticism, European Romantic Review,* and *ELH* and has held a Barbara Thom Post-Doctoral Fellowship at the Huntington Library. Her book *Poetry, Enclosure, and the Vernacular Landscape, 1700–1830* (2002), explores the shift from the values of the open prospect to that of confined spaces from the Act of Union through the Regency period.

NOAH HERINGMAN is assistant professor of English at the University of Missouri–Columbia and was an N.E.H. Fellow at the Huntington Library in 2000–2001. His forthcoming book, *Romantic Rocks, Aesthetic Geology,* demonstrates the mutual importance of Romanticism and early geology. He has articles published or forthcoming in *Studies in Romanticism* and *Huntington Library Quarterly.*

THERESA M. KELLEY is Marjorie and Lorin Tiefenthaler Professor of English at the University of Wisconsin–Madison. She has published *Wordsworth's Revisionary Aesthetics* (1988) and *Reinventing Allegory* (1997), as well as essays on Romantic poetics, aesthetics, rhetoric, textual studies, and the sister arts. *Reinventing Allegory* was awarded the 1997 annual SCMLA prize for best scholarly book. She also edited (with Paula Feldman) *Romantic Women Writers: Voices and Countervoices* (1995).

Kelley's current project concerns the role of botany in Romantic culture as a case study in the theory and practice of Romantic aesthetics and culture.

AMY MAE KING is assistant professor of literature at the California Institute of Technology. Her forthcoming book, entitled *Bloom: The Botanical Vernacular in the English Novel*, explores the evolving impact of the natural sciences—in particular, Linnaean botany—on the novel, especially in its representation of the physicality of courtship. A chapter from this project has appeared in *Eighteenth-Century Novel*.

LYDIA H. LIU is professor of comparative literature and Helmut F. Stern Professor of Chinese Studies at the University of Michigan at Ann Arbor. Her books include *Translingual Practice: Literature, National Culture, and Translated Modernity* (1995), an edited volume entitled *Tokens of Exchange: The Problem of Translation in Global Circulations* (1999), and a coedited book with Judith Zeilin, *Writing and Materiality in China* (2003). Her new book *Desire and Sovereign Thinking* will be published by Harvard University Press. Liu has also begun work on a cross-cultural study of porcelain, on which the chapter in this volume is based.

ANNE K. MELLOR is professor of English at the University of California–Los Angeles. Her books include *Mothers of the Nation: Women's Political Writing in England, 1780–1830* (2000), *Romanticism and Gender* (1993), *Mary Shelley: Her Life, Her Fiction, Her Monsters* (1988), and other works on British Romanticism. She has also edited collections of essays and (with Richard Matlak) a major anthology of Romantic literature: *British Literature: 1780–1830*. Her earlier work on literature and science includes an essay on *Frankenstein* in *One Culture*, ed. George Levine (1987).

STUART PETERFREUND is professor of English at Northeastern University. His most recent books are *Shelley Among Others: The Play of the Intertext and the Idea of Language* (2002) and *William Blake in a Newtonian World: Argument as Art and Science* (1998). Peterfreund, a past president of the Society for Literature and Science, has also edited the collection *Literature and Science: Theory and Practice* (1990) and has published numerous articles on Romanticism and natural history.

CATHERINE E. ROSS is assistant professor of English at the University of Texas–Tyler. She is at work on a book project entitled *Rivals in the Public Sphere: Humphry Davy and the Romantic Poets*. This project is based on her dissertation, which won an Outstanding Dissertation Award at the University of Texas–Austin in 1999. Her most recent presentation from this project was a plenary address at the International Society for the History of Anesthesiology convention in Santiago, Spain, in 2001. Her paper will appear in the published proceedings from this conference, and she also has essays forthcoming on the poetry of Samuel Taylor Coleridge and Elizabeth Barrett Browning.

Index